T0145263

Systems Engineering
Agile Design Methodologies

James A. Crowder · Shelli Friess

Systems Engineering Agile Design Methodologies

James A. Crowder
Raytheon
Englewood, CO
USA

Shelli Friess
Relevant Counseling LLC
Englewood, CO
USA

ISBN 978-1-4939-4772-0 ISBN 978-1-4614-6663-5 (eBook)
DOI 10.1007/978-1-4614-6663-5
Springer New York Heidelberg Dordrecht London

Printed on acid-free paper

Springer is part of Springer Science + Business Media (www.springer.com)

Preface

Dr. Crowder has been involved in the research, design, development, implementation, and installation of engineering systems from several thousand dollars up to a few billion dollars. Both Dr. Crowder and Ms. Friess have been involved in raving successes and dismal failures not only in development efforts, but in team building and team dynamics as well. All the failures have a common theme; the inability of engineers, managers, and teams to respond well to change, whether changes were due to problems in the development, or changes because the requirements for the system were in flux. While this certainly was not the only problem, it was a large contributing factor. The resistance to change has been a part of not just engineers, but people in general, since humans first began to create and build. However, the world is changing faster than ever before, and will continue to change, not just at the same rate, but at an ever increasing rate as time progresses. The organizations that survive will be those that have technology, people, processes, and methods that allow for and embrace change as a normal part of doing business.

This book was written to help engineering organizations understand not just the need for change, but to suggest methodologies, technologies, information systems, management strategies, processes, and procedural philosophies that will allow them to move into the future and be successful over the long term in our new information-rich, hypermedia-driven, and global environment.

This book is not intended to be exhaustive, but to introduce systems, software, hardware, and test engineers, as well as management to a new way of thinking; a new way of doing business. This includes not just the technologies and organizational structures, but the team dynamics and soft people skills that will be required to create, attain, retain, and facilitate efficient product teams required for future engineering development efforts. In short, we have to rethink everything we have ever thought about how to design, build, install, and maintain engineering systems. This book is a start along that process.

Contents

Chapter 1
Introduction

God, give us the serenity to accept what cannot be changed;
Give us the courage to change what should be changed;
Give us the wisdom to distinguish one from the other.
Reinhold Niebuhr, 1892–1971 [1].

1.1 Change as a Precept Rather than a Fear

The agile design methodology has become an unavoidable factor in the modern design paradigm. The formal agile software design process has been utilized since the mid-1990s. Scrum was introduced in 1995 and Extreme Programming in 1996. In 2001, the Agile Alliance was formed and established the Agile Manifesto [4], which states:

- *Individuals and interactions*: over processes and tools.
- *Working software*: over comprehensive documentation.
- *Customer collaboration*: over contract negotiation.
- *Responding to change*: over following a plan.

There was, and still, much resistance to agile software development. However, it is an accepted, modern method for development of software in the new world of rapid development. Unfortunately, systems engineering design methods have not kept pace with software, creating schisms between the systems designs and the final delivered software. A new culture must be created that integrates research and development (R&D), systems, software, test, and reuse into a modern, agile, methodology that makes change the rule or order, rather than the exception. Agile engineering is a conundrum of fact, science, folklore, and misconceptions. This drives the need not just for wisdom, but for well-established methodologies in all areas pertaining to engineering design. Three areas of change will be discussed here that are critical to creating an overall agile design methodology.

- New organizations are required to take advantage of new technology environments. Examples are:
  - Domain centric
  - Design teams
  - Concurrent Engineering
  - Systems of Systems Enterprise Architectures

J. A. Crowder and S. Friess, *Systems Engineering Agile Design Methodologies*,
DOI: 10.1007/978-1-4614-6663-5_1, © Springer Science+Business Media New York 2013

- New technologies for harnessing communications and automated design methods, like:
  - Common tools and modeling techniques for:

Systems engineering
Software engineering
Human Systems Interfaces and Controls
User Engineering

- New methodologies for increased quality, reduced waste, and with reuse and test built into the overall process:
  - Quality Function Deployment (QFD)
  - Continuous processes improvement
  - Software Cleanrooms

The new organization addresses the need for a transition to new organizational team dynamics that go beyond Integration Product Teams (IPTs), utilizing software and Information Systems (IS) and Knowledge Management (KM) technologies to facilitate seamless collaboration between systems, software, and test; including the use of process design automation tools like QFD to facilitate robust, agile designs, and a new concept, called Functionbases, which capture the systems, software, and test design, as well as the context for the design in one hypermedia reuse package.

New technologies' effects on Commercial Off-the-Self (COTS) software will be reviewed with an emphasis on hypermedia concepts to facilitate correct and agile systems, algorithm, software, and test designs. Ensuring quality is related to robust design algorithms that are required to meet and guarantee performance not just of the final product, but performance characteristics of the overall design process as well.

1.2 The Historical Significance of Change

Change has become an unavoidable factor in the modern design process. This is partly due to constantly changing computer technology and its effect on the design process [7]. The effect that IS technology is having on the engineer's tools is understood. The tools are providing a revolutionary effect on the organization of the engineer's work, but what must also be understood are the cultural changes needed in the design process to match the evolution of the tools [23]. Without this understanding we become outmoded, our relative productivity drops, and our engineering becomes increasingly uncompetitive.

Revolutions in human development yield equally revolutionary increases in human productivity [12]. The first was the Agricultural Revolution in 6000 B.C., thought to have been brought about by changing climatic conditions. It was farming that spawned so much of civilization that we now take for granted, especially the accumulation of wealth and the forming of a government to protect that wealth. The second was the Gunpowder Revolution in the fifteenth century and the development of the cannon. This revolution gave rise to a new military technology and a centralization of power that led to higher levels of organization and greater wealth development, specifically the Industrial Revolution. The third revolution is the Information Revolution (which we are in now), where the raw materials for wealth development are no longer energy, ore, and muscle, but computer technology, data, and intellect. The Information Revolution should increase productivity through higher quality work (and workforce) to reduce waste.

The trouble with implementing change in the engineering process is that it is bound to upset someone. It is, in fact, akin to being an unwelcome prophet. However, avoiding change is hazardous, and often so on a grand scale. The following examples illustrate the point.

1.2.1 The Stirrup

The invention of the stirrup meant that owners of horses had a powerful weapon that could not be defeated by a soldier on-foot. The stirrup afforded the rider greater stability and, therefore, greater lethality. Because horses were a commodity in short supply, this new form of power was soon monopolized. The result was a new social structure: feudalism. In this structure, the knightly class owned the horses, and hence the power, the peasant class supplied the materials. In fact, the word "imbecile" has its origins in the Gothic Latin word imbelle, which originally was applied to the scorned masses of peasants who did not own horses and were consequently weak [3].

Failure to implement this new technology explains King Harold's loss of the Battle of Hastings in 1066 [39]. If King Harold had possessed the same well-armed knights as King William, the outcome might have been quite different. King Harold suffered from being on an island isolated from technological change in Europe.

1.2.2 The Luddite

Early in the nineteenth century, new labor saving devices invented in England enabled greater productivity in the textile industry. Ned Ludd and his followers disapproved on the grounds that it put at risk their livelihood by diminishing their employment (the loom would replace them all). Their solution was to tour the countryside destroying any new loom they could find. This solution was short-sighted as change was inevitable. The term "Luddite" has since become part of the

English vocabulary to identify anyone who opposes new technology on the grounds that it is labor saving.

The moral is simple: King Harold's folly was being isolated from change, and the Luddites resisted change.

By ignoring the impact of IS and KM on the engineering process, we will be repeating the folly and never know until it is too late that we are no longer competitive in the marketplace.

1.3 The Modern Design Folly: Engineering Processes and Metrics

Having been involved in systems engineering of 25 years for most of the major aerospace companies, the following has been often overheard from program managers complaining about the same things:

Why is our overall productivity the same as it was in the 1900s?

Why is our cost of test exceeding the cost of development?

There have been many advances in software development over the past 15 years that have sought to dramatically improve the efficiency of code production. However, similar advances have not been seen in systems architecture and test design to keep pace with agile and extreme programming initiatives. And while the industry pushes to utilize Object Oriented System Engineering, with tools like SysML,[1] and Test-Driven Development [5], we can no longer see system engineering, software development, and testing as separate organizations and separate entities within a development project. Today we are faced with market demands and evolving computer/IS technologies [15]. The modern equivalent to King Harold's Folly and the Luddite mentality is the reliance on "our process" and "its associated metrics." Too often we rely on these processes and metrics to form the bases for our proposals. Treating each of these design/development elements as separate is why we still struggle with overall program execution productivity, are often faced with costs that exceed projections, and end of with a final system that does not conform to the design requirements and expectations.

As the computer and IS technologies continually improve, and our development/design tools become more sophisticated, we tend to apply these improved capabilities to enhance our proposals; however, the underlying processes and metrics are utilized to form the basis of the bid. More often than not, the technologies are applied to current processes, which are then utilized to generate the same kind of metrics. What we engage in is folly because we fail to realize that these new technologies allow for a shift in engineering methods (e.g., automated

[1] http://www.omgsysml.org/

design, test, documentation, etc.), which changes productivity, hence, change the metrics that form the bases for our bids.

We have seen over the past decade that computer, IS, and collaborative technologies continually improve exponentially and appear to provide the abilities to improve long-term productivity. However, long-term productivity has severely failed to match the advances in technology. Many theories have been posed to explain this phenomena; more complex problems, increased customer demands, increase in data, etc. We see that one of the major long-term problems affecting engineering organizations is "metric inflation." This is caused by our continued reliance on "tried and true" metrics, most of which are ineffective and wrong for modern systems design/development/implementation, hence causing long-term problems with our business models.

For most organizations, program performance is measured against the success of meeting classical metrics (e.g., cost, schedule, line-of-code counts, etc.). Therefore, if the proposal calls for 300 staff positions (systems, software, test, hardware, etc.), these positions are staffed, based on these classical metrics. This occurs without a complete and full productivity assessment against improved computer/IS/collaborative technologies, but is based on the same tried-and-true metrics we have been using since the 1960s. This becomes what we will call a "self-licking ice cream cone," in that the application of improved productivity tools is offset by conventional staffing metrics.

Inflation occurs by a number of factors, examples of these inflation factors are:

- Rather than modification of the engineering processes to account for new technology advances, more "process" is applied to the engineering methods and its use of technology to "improve" the process, therefore decreasing the efficiency of any improvements and actually decreasing the expected increases in quality, since too much process interferes with the creating process and mistakes are made. We fall too easily into the "Process over Productivity" paradigm.
- Budgets are always fully expended; therefore, work is expanded to fill the budget and schedule allowances.

Over the long term, over the course of program execution, any savings that might have been felt (which equate to improved competitiveness) in the form of improved productivity, are hidden beneath overbearing processes and budget expenditures. For an organization to remain competitive, the engineering metrics must be modified to baseline them against improved productivity tools and technologies. This is not to say that process in and of itself is inherently bad. Many a program has failed because of lack of any engineering processes. However, it does not follow that more process makes the project better and more productive. There must be a balance. And the processes that are used must be appropriate for the project. One size does not fit all when it comes to engineering processes; and that does not mean tailoring out given sets of processes for small, medium, or large programs. It means having the right processes and the right metrics for each individual project.

The engineering methods and processes must evolve to account for new technologies and methods, and the metrics associated with these changes must also be evolved and established in order for an engineering organization to remain competitive. The ability to evolve and remain competitive requires engineering organizations to challenge its processes and metrics constantly.

1.4 The Modern Design Folly: Embracing Modern Capabilities

As discussed above, one of the major challenges for engineering organizations is to remain competitive in the marketplace [26]. In the past, the capabilities of computer technologies restricted the project solution space to the developers and consumer (customer). In today's modern, ever evolving and improving environment, computer/IS/collaborative (communication) infrastructures have enabled massive expansion in available services and capabilities [11].

With the integration of automated design and development tools available to all elements of the design/development/implementation/execution phases, the design of the system has become more important than the physical development, meaning we can now:

- Design a new enterprise design systems, and allow the automation tools to validate the design, and produce the documentation to support the design.
- Allow the automation tools to build software bases, based on the validated application/service design.
- Design portals, and their support displays, which allow the automated development tools to build and test the software.
- Design data management systems which allow the automated development tools to build and test the data management software.
- Design and build automated testing tools for unit, element, subsystem, system, internal, and external interfaces; which then enables continuous automated integration and regression testing.

Modern computer and IS technologies are affecting our engineering methods, processes, and metrics, and are outpacing the bureaucratic development organizations' ability to keep pace with these improvements [33]. Engineering organizations must be prepared to accept, and embrace, change as a normal part of its business model, in order to remain relevant and competitive in the customer market place.

1.5 Layout of the Book

We have arranged the book to build up to new methods for Agile Systems Development. We start by understanding why people are resistant to change, particularly engineers, and then describe a modern, agile system design methodology that is created to help people develop a philosophy of change as a way of life. The progression of the book is described below:

Chapter 2: This describes why people, and in particular engineers, are resistant to change, and how we can provide them tools to embrace change as a normal part of their design process.

Chapter 3: This chapter describes changes that are required within engineering organizations to facilitate modern design methodologies and processes and what types of organizational structures actually suppress the modern, agile design methodologies.

Chapter 4: Here we emphasize the domains that engineering organizations must master in order to promote and execute Agile System Design methods.

Chapter 5: This chapter describes the types of organizations that are required for modern system design, those that promote productive behavior. We introduce new organizational structures (e.g., consensus engineering that facilitates collaborative engineering) that are required for future design methods. Also included are design methods that will be required to drive agile systems designs, and the transitions needed to get there (e.g., eliminate stove-pipe engineering).

Chapter 6: Here we introduce the new Informational and KM techniques that provide the capabilities for agile systems designs, including a new automated design tool, called the Functionbase, which provides a complete agile systems design process capture and reuse paradigm.

Chapter 7: This chapter discusses the total agile systems design process, including tools, that allow major increases in quality and efficiency over current design and test methods.

Chapter 8: Here we wrap up our discussion, again emphasizing the need to embrace change, and how the methods discussed in the book more easily allow engineering organizations to embrace change as a normal part of their everyday existence.

Chapter 2
The Psychology of Change

If you do not change your direction, you may end up where you are going.

Lao Tzu

2.1 Why Do We Resist Change?

If change has always been an integral part of life, why do we resist it so? Why, in every generation, do we have Luddites? What goes through a person's mind when they are informed of or predict change? Is my position safe? What will I have to do? What will I need to know? Am I capable and confident with new direction? Do I have any say about this or any control over what is about to happen? Do I really need to change? Do I have time for this? How am I going to do that and this? How will this impact what I have already done? As you can see there are many questions that come up when even the thought of change occurs. It seems obvious that there should be thought put into change theories as we ask for people and environments to change.

Some key components to encourage change are empowerment and communication. People need time to think about expected change. As we discussed earlier, people are good at change in order to master or improve their world or environment. When people remain agile they are better at being agile. When change is part of the regular process then one becomes agile. Alternatively, when one is used to doing something exactly the same way or systematically then it becomes the way one likes to operate.

Why, in particular, do some engineers not like change? When asking this question it seems obvious to consider education. What is it that has been required of engineers in the past and present and what will be required of them in the future. Lucena [40] hypothesizes connecting engineers' educational experiences with their response to organizational change and offers a curriculum proposal to help engineers prepare for changing work organizations. As our technology increases and our work world becomes more agile it makes sense also that soft skills will become more and more important for engineers. Engineers can organize themselves to optimize performance with soft skills.

Trust is the most important factor in change. Trust helps to balance fear which is often the root of most resistance. Who in their right mind will blindly make changes without trusting others? Agile methods increase trust by increasing transparency, accountability, communication, and knowledge sharing [41]. Iteration/sprint planning

J. A. Crowder and S. Friess, *Systems Engineering Agile Design Methodologies*,
DOI: 10.1007/978-1-4614-6663-5_2, © Springer Science+Business Media New York 2013

methods give members visibility on requirements, individual assignment, and agreed estimates. People get the information at the same time. The daily stand up provides visibility and transparency so issues can be addressed immediately. People will know if someone is behind. The sprint/iteration retrospective provides transparency and visibility regarding goals. This agile design builds trust among the team members as they are able to see others' trustworthiness and competencies as they continue working together.

Crawford et al. [14] suggest that employees who had higher creativity, worked on complex challenging jobs, were supportive, had supportive non-controlling leadership, and produced more creative work. If workers see that their ideas are encouraged and accepted they are more likely to be creative. Empowerment and trust encourage creativity which encourages change which encourages agility [41].

Think about the complexities that people bring to projects. Think about projects of many teams made up of many people, virtual teams, many sites, many locations, many levels of expertise and experience, many cultures and places that people live. Then add in technology, many technologies, changing technologies, developing, and revolutionary technologies. It becomes more and more clear that soft skills are incredibly valuable to engineers and particularly to agile teams. Azim et al. [2] show that up to 75 % of project complexity has to do with the human factor or the people in the project. They claim that soft skills are important in the implementation of plans. Soft skills are clearly important in any complex project and in all phases of change previously discussed, particularly in the commitment and transition of change.

Soft skills include organizational, teamwork, communication, and other people-based skills. As technology matures and new technologies emerge it is imperative that the teams and people in the teams become more agile. Not only is technology constantly changing, so are people. It seems ever so important that the soft skills come with the hard skills.

One of the main thrusts of this book is to introduce the reader to a new organizational structure, called the Theory Z organization (Chap. 5) in order to facilitate a system engineering design methodology that embraces and handles change as an integral part of the normal process flow. We will refer to the following discussion throughout the book, in reference to Theory Z management and organizational structures. We will briefly introduce Theory Z here with much more detail to come later.

Theory Z: Described as "consensus decision making," establishes strong bonds of responsibilities between team leads and team members, with a high importance placed on finding people with the right skills, both "hard" skills (e.g., technical) and "soft" skills (e.g., creative thinking) for team creation. This creates a seven-point structure for Theory Z teams:

1. Strategic plan of action.
2. Strategic team organization.
3. Systems/software/test/hardware formal and informal procedures that support the strategy and structure.
4. Team goals—guiding concepts.

5. Staff—people/human resources.
6. Skills—the right skills, not generic engineers.
7. Style—soft skills that facilitate cooperation and collaboration within the team: the cultural style of the organization.

2.2 How Do We Embrace Change?

There are many psychological considerations to think about in relation to Theory Z structures. First, let us consider humans and their social interests, tagged by Alfred Adler as *Gemeinschaftsgefühl* [8], which described people's state of social connectedness as part of their overall psychological health. Let us assume that all healthy humans have these innate social interests in any given environment. Thus, people can create group goals and together work toward a common goal or product as a group. One could interpret Adler's theory as saying people have the will to change for the better, the will toward mastery of problems, or the will toward perfection. People desire to change, to grow, to overcome, to master. They have the creative power to redirect efforts. This is very different from a will to power, different from directing the group to bettering the group. In a group of healthy people, the group as a whole is likely more satisfied with working toward a common goal of the group than working toward being the power of the group. Given the Type Z structure, people have the potential to be more successful than the top-down management styles. People can work together to address changes and solve problems better than being told what to do and stick to specific tasks. People are adaptable, not static, but ever changing.

Of course, context is always important to consider. There are some contexts that will require different roles to be implemented. There is a time for some direction to the group and there is a time to get direction from the group. Specifically, each member plays an important role while at the same time so does the group as a whole. Thus, there is a time for facilitation within the group and a time for facilitating between groups. As stated previously, this is the structural difference with Theory Z.

It is important when facilitating groups of people to consider group dynamics. There are different theories about groups that have been well documented. According to Jacobs et al. [34] all groups go through three stages: the beginning, the middle or working stage, and the end stage. Briefly stated, the beginning stage is about getting to know and be comfortable with one another. It is a time to define goals or problems that are going to be addressed. The middle stage is the working stage where things happen and the group can see itself as a whole or mostly whole.

At this point the group is productively working to accomplish its goals. The end stage is the termination point where people can consider their accomplishments and what they were able to get from being in the group. Most engineering organizations subvert this process by creating Integrated Product Teams at random, filled with a mix of generic skills (H/W, S/W, systems, test, etc.) and assume, by sheer force of management's will, they can be made to be a productive team.

Tuckman [51] talked about developmental stages of groups depending on their group activities. Relevant to our discussion in this book is his work in describing task-oriented interpersonal groups as going through stages such as testing-dependence, conflict-resolution, cohesion, and functional role determination. In particular, Tuckman concentrated on the task-oriented group and their development phases: orientation, emotionality, relevant opinion exchange, and emergence of solutions. Tuckman addressed the separation of interpersonal groups from task groups in that each person brings their self into the task group as well as the social group (note, a development team is a form of social group). People may not be sharing about personal issues in a task group but their *personhood* will show up. Nonetheless, people are going to follow the dynamics that it takes to get to the working stages of the group. They will have to figure out what role they play, what are the norms of the group, how safe it is to provide their input, do they trust others, how will the facilitator respond, etc. Groups go through these phases, termed forming, storming, norming, and performing. The facilitator's role is to get the group to the performing stage. The quality function deployment's (QFD) role would optimize productivity while paying attention to group dynamics. We will discuss the QFD in detail later in this book.

In "Forming to Performing: The Evolution of an Agile Team", the authors suggest that the stages of groups are cyclical in nature [45]. Tasks that are and are not accomplished in one phase may be revisited in others. Yet, they say, what will remain constant are the agile principles. Thus, supporting the role of the QFD to facilitate and promote necessary changes within the group to reach the optimal performance of the group. Relating back to Tuckman [51], these authors talk about different tasks they identified throughout their group processes. In the beginning they talk about testing tolerance, working style, providing direction, and unclear processes. They relate that during the storming phase the group was characterized by polarization, personal agendas, direct guidance, attrition, and factions. The norming phase consisted of cohesion returning and fostering collaboration. In the performing stage they were able to identify total focus, collective decision making, collaboration, and productivity.

To consider Theory Z is to consider change and to consider people's response to change. We can see thus far that people like change as it comes to mastering or bettering themselves and thus their product. So why is it that many people seem to resist change? Let us consider some reasons why we might see this. First and foremost is fear. People do not like change if they fear losing something they value. Secondly, people prefer that their world and environments are predictable, so that they can know what to expect. It makes sense that people are on edge in unpredictable environments because they are spending more time in trying to figure out what is going to take place. This is where Theory Z may be helpful in that when people are part of the decision process they are more likely to see it as predictable or at minimum spend less time worrying about what is next. They will also be able to find comfort with the group being involved in decisions because there is less guess about where management came up with the decision to operate in that way to get some end result. People experience change often. The facilitator or QFD can promote healthy change that keeps stress to a minimum.

2.3 Components of Change

In "A Meta Model of Change" [55] the author writes about a Meta-Analysis of many change theories. He found nine common themes that are appropriate for our discussion. He writes that change starts with an *existing paradigm* and the nature of this paradigm will determine if there will be recognition that change is needed. Next, there is a *stimulus* and then *consideration*. The stimulus is the motivator and the consideration is limited by the observer. He also observes that there is a need for different viewpoints. The next stage involves what he describes as *validating the need*. This stage answers the question: is there enough evidence that change should occur? Once it is determined that change should occur then one must *prepare*, plan, or reengineer. The following step is a *commitment to act* followed by what he calls *transition;* the *do-check-act*. Here we ask: is the vision or reengineering meeting the goals and do adjustments need to be made? The next phase defines the *specific results* of the efforts to change. The final phase is *enduring the benefits*. The change has also produced the ability to change and all that comes with the finished product.

2.4 The Encouragers of Change

Some of the strongest encouragers of change are trust and empowerment. Empowerment allows people to develop and make decisions with confidence. Trust is very similar. Trust and empowerment go hand in hand. These give way for humans to make change that results in more change.

In a general sense, empowerment can be thought of as a motivator. It is about living up to one's potential. It includes making choices and taking actions. Wan and Zhou [53] suggest that when an individual feels "authorized" they may change attitude and behavior. Empowerment gives people a sense of authorization. How differently does one act when they feel authorized versus told what to do and to get something done? Honald [31] writes that each organization must define exactly what it is that they mean by empowerment and that empowerment must address the needs and culture of each unique entity. How does each entity give authority to or afford people to meet their potential? If we want people comfortable with change then we need to empower them to have authority to make decisions and have choices upon what actions to take.

How many more choices will an individual see when they feel empowered versus ordered. Those are extremes but it helps bring the point home. Empowerment gives people a sense of ownership and an opportunity for creativity. People can come up with new ideas and solutions to problems and are more likely to be motivated to work toward those solutions when they have some personal investment in them.

Chapter 3
The Modern Design Philosophy: Avoiding Change is Perilous

The worst thing that can happen to this organization is to receive another 100 % award fee.

Program Manager, circa 2000.

While most upper management would probably get pitch forks and torches and come after a program manager for such a blasphemous statement, the simple fact is that human nature is what it is: if there is no stimulus that drives us to change, organizations continue to conduct business as usual and ignore opportunities to grow, evolve, and change. The following was actually spoken by a customer to an organization:

> You are a dinosaur that is going extinct, you just don't realize it and won't until you're gone.
>
> Government Customer, circa 2002.

3.1 Organizations that Suppress the Agile Systems Design Methodology

Why are engineering organizations so resistant to change? Historically, engineering organizations have been designed to resist change [29]. Change has been generally considered undesirable, as typical thinking believes that change leads to increased cost and increased uncertainty (or risk), especially when the product is new. This engineering philosophy, what we think of as a "risk averse" philosophy, has led to the development of the classical theory for engineering organizations, and has driven system engineering design for decades. This methodology was forged by an integration of the scientific management method and the ideal bureaucratic method.

The scientific management approach was principally developed by Fredrick Taylor in the 1900s [48]. Taylor's objective was to remove chaotic change from the engineering and manufacturing process, thereby increasing corporate wealth for the enjoyment of all (both worker and manager). His approach was to scientifically analyze the process so as to define the best work methods and worker skills for the tasks. It was much more than time and motion studies; it was an early example of process management.

J. A. Crowder and S. Friess, *Systems Engineering Agile Design Methodologies*,
DOI: 10.1007/978-1-4614-6663-5_3, © Springer Science+Business Media New York 2013

The ideal bureaucratic method is associated with Max Weber [54] and the objective was to define the best types of organizations to active company goals. His method was to design job descriptions and a supervisory hierarchy that was optimal. This led to job specialization and rigid structures with their respective shortcomings that we often associate with the term “red tape”. Such a system is most effective for invariant processes, most of which have nothing to do with long-term development projects. This type of management and team structure is wholly inadequate for a dynamic development process. Unfortunately, the bureaucratic organization is the most prevalent, as it is easy to conceptualize. Even more unfortunate, most organizations consider bureaucracies essential to manage changes in a complex system, due to the risks involved. They often add process on top of process, review board on top of review board, assuming more process will reduce risk and enhance engineering. Neither are true, as change is inevitable, and change is only risky when managed by a bureaucratic organization. This is not to say that we should not manage change, however, the ability to handle change should be an integral part of the overall design and the overall organization, and not a bureaucratic methodology.

Many modern engineering organizations have tried to move to, what is thought to be, a more suitable organizations structure, the Matrix organization. This organizational structure manages both the product structure and the functional structure simultaneously, in an attempt to create the “best of both worlds” for engineering. However, in most cases, this creates a struggle, as each engineer has two supervisors and a confused chain of command, as it is not unusual for both structures to be in disharmony. But, the Matrix Organization method is, at least, an attempt to manage change positively.

3.2 The Push for Continual Change: Agile Design Patterns

> We have gone from an environment where change was seen as a source of risk to our customers, to a new environment where organizations that cannot adapt to change being seen the new source of risk.
>
> Dr. James A. Crowder (2010) AIAA Conference.

3.2.1 The Lack of Ability to Change is the New Risk

When you consider the technological changes that have happened over the last 40 years, and more importantly, over the last 10 years, you see an exponential increase in capabilities, not just in computing power, but in Human System Interfaces (HSI), network technologies, wireless technologies, miniaturization technologies (e.g., nanotechnology), energy technologies, etc. For large, multiyear development projects, if an organization has not developed technologies, organizations, procedures, and methodologies that allows for rapid change in requirements, designs, and technologies, they risk having a final product that is no longer relevant for environments and requirements it was designed to meet.

In addition, not only does technology change at an alarming rate, sociopolitical needs and human needs in general change at a pace such that development projects that run over 5–10 years will find it virtually impossible to bring a product to market that is still needed or relevant; unless the development team (including hardware, software, systems, test, and management) is equipped and adapt at transforming the development, mid-stream, to account for all of the changes it encounters along the development path. This drives the need for new "soft" skills that are needed to deal with engineering development in an environment of change, for change is not constant, change is every increasing.

3.2.2 The Velocity of Change is Increasing

In her article for the Daily Record, Nicole Black discussed the incredible speed at which technology is advancing [7]. Her conclusion is that technology is advancing faster than the legal and social communities can keep up with and ethics committees and lawmakers need guidance from technology pioneers just to understand and navigate the ever-changing technology world we live in.

As the velocity of technology increases and the need to have organizations that are not just adapt at change, but are geared to embrace and use change as an integral part of their development, we have to move away from classical management structures. If we look at Matrix Management and its attempts to manage employees, careers, projects, and functional needs, we find an organization that is antithetic to change. In terms of "managing for change", the classical Matrix Management would be considered an "Anti-pattern". Our current Integrated Product Team structures are constructed with people from a variety of disciplines, each of which has a separate matrix manager, possibly a functional manager from within the project, an IPT lead, and a project manager. There is no team cohesiveness because every member of the team has cross purposes driven by a multiorthogonal organizations structure.

The next problem with classical matrix and IPT organizational structures is that people are not trained with the "soft" people skills to adequately manage an actual integrated team, where interpersonal team dynamics becomes extremely important for success. Most organizations try to "fix" this problem with engineering directives and multiple engineering processes that do not drive engineering, but hamper and kill creativity, making the team a danger to change. That is not to say process is not important, but what is needed is processes, technologies, and methodologies that help manage and embrace change. Also, classical program management training does not teach the soft people skills necessary to facilitate teams of agile systems developers. These include skills like, communication, conflict resolution, negotiation, personal effectiveness, creative problem solving, strategic thinking, and team building. All of these are critical for development teams to be effective, especially in an agile development environment where change is the rule, not the exception. Obviously, a development team must also have the technical skills to accomplish the tasks, but just having the technical skills will not create a successful development team.

The rest of this book is devoted to providing a roadmap for new organization structures, new hypermedia development tools, and new methodologies that will be required to accommodate development projects in the future.

Chapter 4
Modern Design Philosophy: Organizational Considerations for Agile Systems Design

In reality, the bureaucratic method is as old as history. Moses, the Lawgiver, implemented a bureaucracy at the suggestion of his father-in-law.[1] Moses was swamped by judgment requests and had to establish a rigid organization to handle the workload. Knowledge was passed down the chain of command in the form of precepts that could be applied to all situations. Because the environment was unchanging, the system worked efficiently and correct behavior was dictated and guaranteed.

Integration of continual change into complex engineering will clearly falter given a bureaucratic organization designed for a constant world. What is required is an organization of loosely coupled product teams[2] that can adapt to changing situations in a stable, well-behaved manner.

The objective is to design a set of job skills that can be combined with information technology to form a profit-center. Use of hierarchy, chain of command, and superior-subordinate dependencies is reduced to enable the team to modify its own organizational structure for optimal performance. The management task is to design an organization of coupled product teams, provide them with their objectives and funding, then monitor them from a distance (in modern terms this is called empowerment). In terms of revolution, the organization is structured for continual change.

The use of the product teams described here broadens the scope of the engineer's authority. This enables autonomous adaptation, minimizing dependency on management. The resulting reduction in information flow upward and commands downward enables a flatter organization that can move faster as the environment changes.

---

[1] The Bible, Old Testament Bureaucracy, Genesis 47:23.

[2] This name implies all the necessary ingredients for product development have been carefully selected such that the team is adaptive to a changing environment.

J. A. Crowder and S. Friess, *Systems Engineering Agile Design Methodologies*,

DOI: 10.1007/978-1-4614-6663-5_4,

4.1 Three Developmental Domains to Master

Advances in computer, information, and design technologies are affecting communications, computing power, and productivity so significantly that market demands are also changing. There are three engineering domains addressed in this book, each illustrated in Fig. 4.1. Each domain has to be addressed to define how the engineer's world must advance to keep pace with the market demands and improving technologies.

First, a new organization must be defined that takes advantage of improved communication methods and the changing market. Computer, network, and wireless technologies are enabling vast amounts of data to be generated and shipped across the globe effortlessly. The large organizational hierarchies designed to collect and sort data for management are becoming redundant, causing organizations to become fatter [3].

Second, new computer technologies are affecting engineering methods. The move to individual workstations enabled stand-alone work environments to be created for individual engineers, with engineering artifacts and data often stored separately. With improved network and wireless connectivity, however, there is value in turning these personal environments into interpersonal collaborative enterprise environments. This becomes more important as the technology increases, enabling more data to be generated and made available, significantly increasing the pace of design. With the constant improvement in computer technologies, and the associated effects on the work environment, changes in design cultures are needed to encourage engineers to work as a team and not as a set of individuals.

Third, increasing quality without increasing cost is possible, as computer and information systems technology can dramatically decrease what was expensive

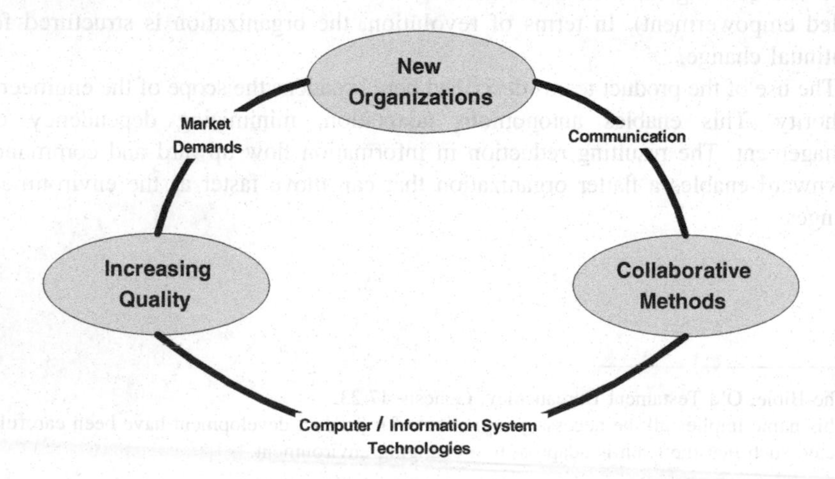

Fig. 4.1 Three developmental domains to master

design, development, and analysis phases of a project.[3] Extensive design, numerical analyses, and simulations can now be run to measure quality before final systems are completed, or even started. This can be used to address demands for improved quality by changing the approach to systems engineering methods (e.g., design, analyses, documentation, and test).

This book addresses each domain in turn. Within them there is a common thread concerning quality: numerous buzz words float around in the market place like "Concurrent Engineering" and "Quality Function Deployment" and will be addressed in detail. They are, in face, attributes of sound engineering and should not be discarded simply because they have been abused as "buzz-words." These are introduced in context to illustrate how a holistic approach to engineering design will capture each of these concepts together to form a fluid, adaptive, and quality/test driven design philosophy that can take engineering forward into and through the new information era of the twenty-first century and beyond. The next chapter will discuss a new organizational structure, beyond Integrated Product Team (IPT) that is required for developmental teams of the future.

[3] Notice we are not invoking "More's Law" as we feel the need/want for information and knowledge always outpaces More's Law.

Chapter 5
A New Organization is Needed: Beyond Integrated Product Teams

I have not the slightest hesitation in making the observation that much of British management doesn't seem to understand the importance of the human factor.
Prince Charles, Speech to Parliamentary Committee, February 1979 [1].

5.1 Organizing for Productive Behavior in the Presence of Change

5.1.1 The Approach

The phrase "throwing good money after bad" can be all too relevant when developing new engineering. Although it is possible to bring new technologies and concepts to an organization, there is no guarantee that they will be implemented. All too often, a new machine is bought only to collect dust.

The starting point is to understand how the engineering organization affects behavior and productivity. Then we can begin to define an organization designed to respond to this new environment of continual, rapid change.

Classical methods of organization did not account for behavioral factors. This was shown with the Hawthorne experiment [27] where worker productivity was measured under different working conditions. Productivity rose whenever the conditions changed, whether for better or worse. The conclusion was that productivity was also tied to how the workers were responding to the sudden interest shown in them. Behavior plays an important role.

5.1.2 Likert's Systems

For the purposes of our discussion here, we will focus on Rensis Likert's research. His contention was that productivity would best benefit from developing work groups that had challenging objectives. Likert defined four systems that describe the span of organizations, from the classical viewpoint that engineers are a cost to be controlled, to a modern viewpoint that considers the engineers to be a resource to be developed.

J. A. Crowder and S. Friess, *Systems Engineering Agile Design Methodologies*,
DOI: 10.1007/978-1-4614-6663-5_5, © Springer Science+Business Media New York 2013

The following is our interpretation of the descriptions of Likert's Systems as described by Hersey and Blanchard [24].

System 1: Management is seen as having no confidence or trust in subordinates, since they are seldom involved in any aspect of the decision-making process. The bulk of the decisions and the goal setting in the organization is made at the top and issued down the chain of command.

System 2: Management is seen as having condescending confidence and trust in subordinates, such as a master has toward a servant. The bulk of the decisions and goal setting of the organization is made at the top, but many decisions are made within a prescribed framework at lower levels. Communication flow is primarily down the hierarchy.

System 3: Management is seen as having substantial, but not complete confidence and trust in subordinates. Broad policy and general decisions are kept at the top, but subordinates are pitted to make more specific decisions at lower levels. Communications flow both up and down the hierarchy.

System 4: Management is seen as having complete confidence and trust in subordinates. Decision-making is widely dispersed throughout the organization, although well integrated. Communication flows not only up and down the hierarchy, but also among peers. There is extensive, friendly superior–subordinate interaction with a high degree of confidence and trust. There is widespread responsibility of the control of the process, with lower units fully involved.

- Likert's *System 1* is compared to McGregor's *Theory X* that describes the classical organization where the manager is solely responsible for defining and supervising the job specializations, infusing knowledge from the top. Similarly,
- Likert's *System 4* is compared to McGregors *Theory Y* where success of the organization is dependent upon both the manager and worker alike. Likert's approach is useful as it helps avoid setting *Theory X* against *Theory Y* as simple opposites. The fact is that *Theory X* is most appropriate for organizations where close supervision is required (e.g., kindergarten).

Because the bureaucratic method requires formal rules and procedures in order to function, it is not seen as spanning more than the first three of Likert's systems. A bureaucratic organization can provide the engineer with the esteem associated with the value of a job well done, but it is unlikely to self-actualize the engineer as that requires the manager and engineer to share decision-making in a profound manner.

The concept of Integrated Product Teams (IPTs) was created and fostered in the 1990s as a way to actualize Likert's System 4 within engineering organizations. However, IPTs were created as a way to provide the illusion of autonomy within an organizations Matrix management structure while under careful supervision of new hierarchies put in place [e.g., SEIT (System Engineering, Integration, and Test)].

In order to establish product teams that can be adaptive to change; the members of product teams must be free to make specific decisions, such as tool selection and engineering work plans. This requires, as a minimum, a System 2 organization.

As the organization matures, the product teams must be free to exercise more responsibility and migrate toward a System 4. Eventually the teams must advance beyond System 4; being responsible for their own education and motivation, and their dependency on the parent organization must continually lessen.

These considerations have been integrated into Fig. 5.1 to illustrate progressive improvement in productivity as organizations implement less System 2 and more System 4 characteristics. Argyris has shown how this migration toward a System 4 management structure does, in fact, improve productivity [25]. There is no doubt that given the computer hardware, software, and humanware resources, a System 4 organization will outperform a System 2 organization.

5.1.3 Beyond System 4: Theory Z

For engineering organizations in the twenty-first century, we need organizations that contain at least System 4 characteristics, and also make complete use of both computer/information technologies and well-educated engineers.

There are aspects of Theory Z that this book will not address; such as the integration of social and work life. We will restrict this book to those aspects of Theory Z that can be summarized by defining self-actualization as a natural integration of work and recreation.

To that end, we define some new parameters that extend Likert's System 4 into the Information Age of the twenty-first century and move us toward a new organizational structure (Theory Z).

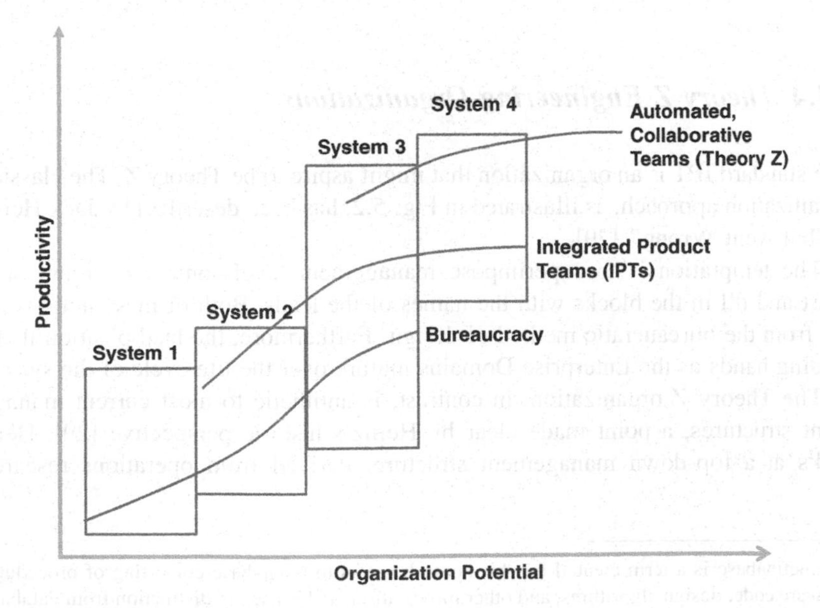

Fig. 5.1 Dependency of productivity on organizational structure

Codeterminism: This redefines the relationship between the manager and the worker, such that the supervisor–subordinate language becomes outmoded. This does NOT mean the elimination of leaders. It means management responsibilities are shared and developed by mutually establishing and allocating objectives. Consistent with this, rewards depend not on individual performance, but on team performance. The method of implementing this will be described later.

Open Functionbases[1]: These enable members of an organization to participate in the processes, which facilitate reuse and rapid integration of new members into teams. This "openness" places an emphasis on continual improvement as it elevates team performance; it is no longer a matter of what you know, but how you use what you know. It is therefore incumbent upon each team member to browse each other's work and to add value for the betterment of the team. The openness is extensible to include geographically distributed members such that the product teams are now "virtual".

Automation of Processes: This, including documentation automation, will transform the amount of spare time an engineer has to pursue continual improvement. Making recurring engineering trivial will transform the work environment such that the larger portion of time is spent on invention and improvement—quite the reverse of today's environment. There will be other advantages in that automation enables the following: recurring engineering will be right the first time, change can be rapidly incorporated, and elimination of wasteful rework efforts. This will vex the Luddite as the workload is diminished. But this attribute of Theory Z is essential if change is to be a natural feature on advanced engineering methods. Quite simply, change takes time, and time must be made available to account for it.

5.1.4 Theory Z Engineering Organizations

The standard IPT is an organization that might aspire to be Theory Z. The classical organization approach, as illustrated in Fig. 5.2, has been described by Jack Heinz, "What went Wrong" [29].

The temptation is to superimpose management levels onto the organization chart and fill in the blocks with the names of the leads. Both of these notions fall out from the bureaucratic method of design. Furthermore, the lead position floats, passing hands as the Enterprise Domains mature over the life cycle of the system.

The Theory Z organization, in contrast, is antithetic to most current management structures, a point made clear by Heinz's historic perspective [29]. Heinz looks at a top-down management structure, derived from operations research,

[1] Functionbase is a term created for this book. It refers to a database consisting of procedures, software code, design algorithms, and other information, and is used in distinction from databases that contain data used by and produced by those functions.

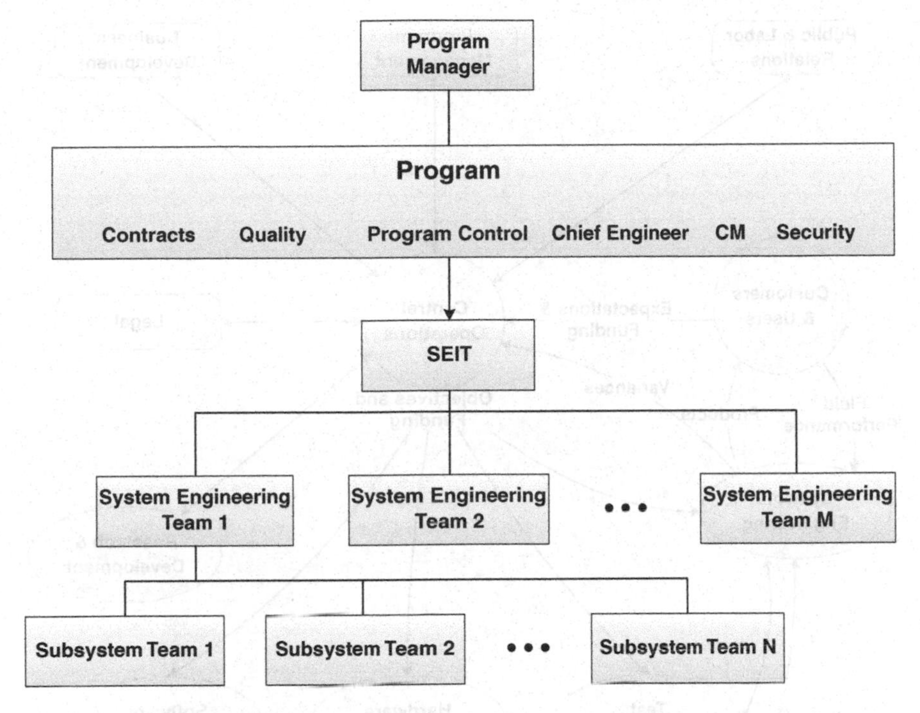

Fig. 5.2 Classical IPT organization

where decisions are made largely at the top. The underpinning assumption is that managers do not need to understand the engineering they are managing. Consequently, expert designers are excluded from the process of managing development, and there is a gradual erosion of the technology base with a consequent loss in productivity.

This loss in productivity affects a growing differential in rewards between the top and bottom of the organization.

In contrast, there is obvious success of several commercial computer software and hardware companies where management is in the hands of competent designers. I would not name them, but you know who they are.

This loss in productivity affects a growing differential in rewards between the top and bottom of the organization.

Instead of the classical organizational structure, an alternative organization, based on Theory Z, is proposed that is based on knowledge workers and an information-based organization. The Theory Z organization, portrayed in Fig. 5.3, is a conceptual model of a human activity system for engineering [12]. As a formal system, it has the following components:

A Mission: to produce higher quality engineering. This renders the systems 'soft' in that it is a continuing pursuit. However, it is also 'hard' in that there are fixed objectives to pursue for each particular product.

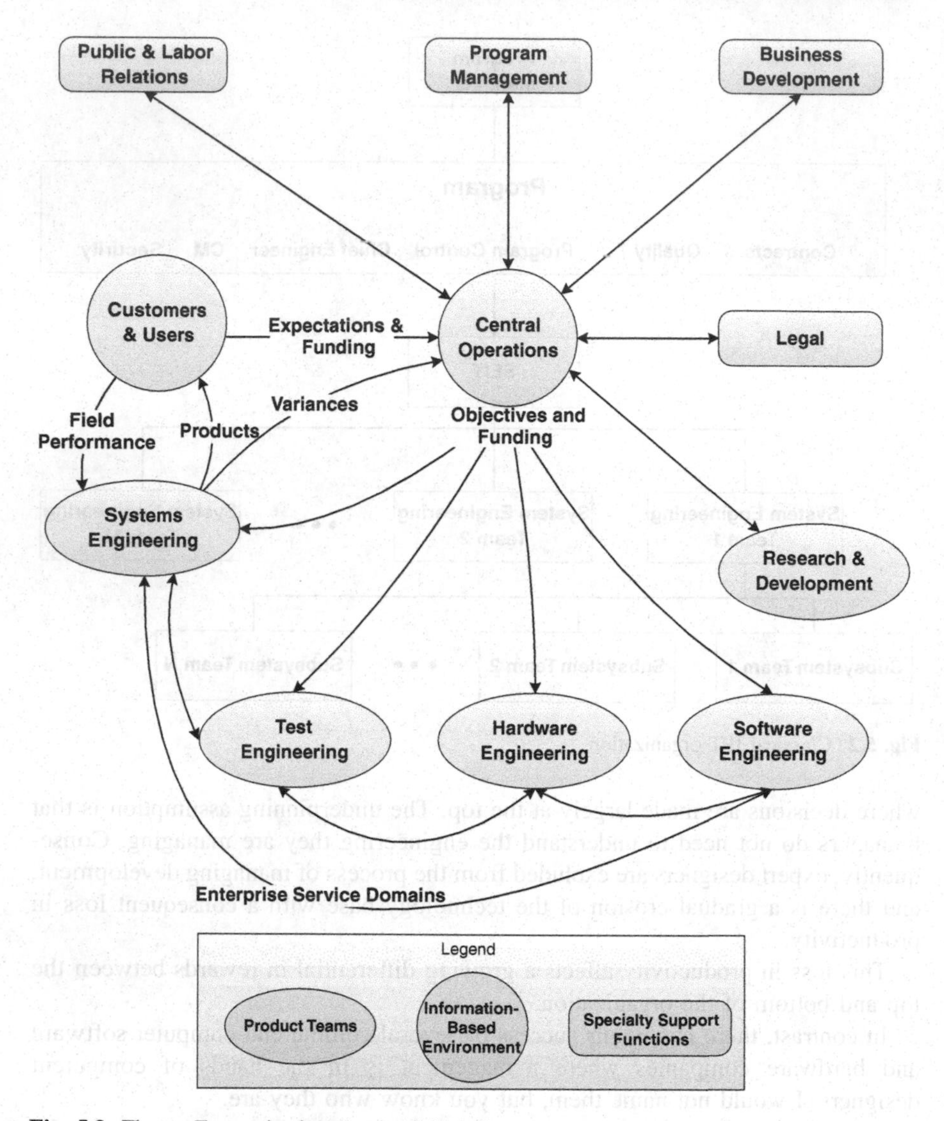

Fig. 5.3 Theory Z organization: responsibility flow

Performance Measurements: to signal process or regress in achieving higher quality engineering.

A Decision-Making Process: that combines the previous two components to provide regulation of the system.

A Boundary: separating the system from its environment. The system can be regulated, but the environment can only be influenced.

The flow of responsibilities is illustrated in Fig. 5.3 in terms of the customer, the central organization, and the product teams. The central theme of the reformed organization is balancing the roles of management and engineering. The Theory Z product teams must be given a high degree of autonomy to practice their skills on objectives established with management, with complete control over their own resources. The Theory Z product team is composed of knowledge specialists who will resist a management chain of command and establish their own informal groups if necessary.

The systems engineering group (SEIT), while not controlling the engineering organization, ensures product integration and collects engineering performance data to compile overall variance information. It is the variance information that is required to derive R&D plans for developing new technology and concepts.

In contrast to the bureaucracy, the reformed Theory Z organization is a flat network of teams. There is no longer a dependency on middle management, because responsibility has been given back to the expert designer to manage the processes locally. This approach is consistent with the Codeterminism and open Functionbase parameters of a Theory Z organization. Another point of detail is the absence of a Quality Department; quality *MUST* be built into the engineering process as will be explained.

Perhaps another way to look at the Theory Z organization is to see it as a learning organization that replicates and improves on the Matrix organizational structure. On one axis there are programs decomposed into product teams responsible for work capital. On the second axis there are corporate critical skills decomposed into their specializations responsible for developing knowledge capital. The term capital is used to describe both axes in order to convey the idea that both work and knowledge are required to make money.

Knowledge capital delineates one company from another and becomes visible during competition. It can be harder to develop than work capital, as it is the union of the knowledge of the engineers within the organization. Figure 5.4 illustrates the union of people within all aspects of a Theory Z organization and Table 5.1 describes the assumptions made by management in the Theory X, Y, and Z organizations.

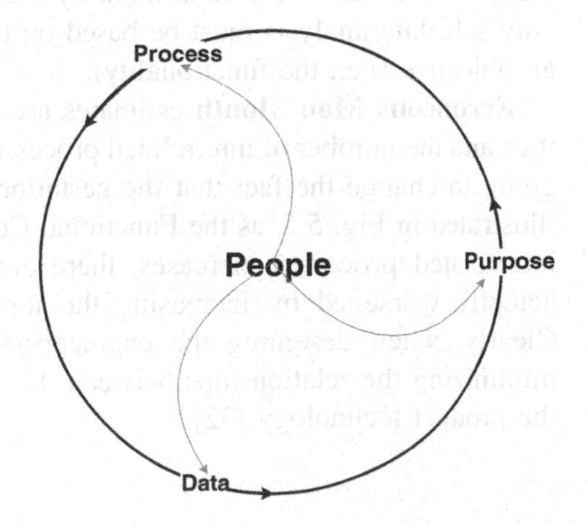

Fig. 5.4 People are central to the Theory Z organization

Table 5.1 Management assumptions across Theory X, Y, and Z

	Theory X	Theory Y	Theory Z
It is human nature to	Avoid work need compulsion shirk responsibility seek to be commanded value security lack ambition	Find work natural if committed: Show initiative Self-control Self-direction Seek responsibility Value creativity	Find work natural if committed: Show initiative Self-control Self-direction Seek responsibility Value creativity
Commitment required	Irrelevant	People need to be committed to the organization	The organization needs to be committed to people

5.1.5 Theory Z Product Team Composition

Imposing discipline from above onto engineering programs with a large, new engineering content is fraught with difficulties. One difficulty has been called the *Mythical Man Month* [9], where planning makes assumptions about how an engineering effort is going to proceed, all of which turns out to be a myth.

Another major problem causing the "Mythical Man Month" is the use of Software Lines of Code (SLOC) to estimate the time required to produce engineering products. In modern engineering environments, with automated code generators and automated processes, SLOC becomes an outdated estimation method, being replaced by estimation methods based on the Function Base of the system (i.e., the number and complexity of functionalities and their interrelated processes).

Fred Brooks, identified this problem back in the 1970s, while managing the OS/360 software program development for IBM [9]. His conclusions can be summarized as follows:

Optimism is a natural attribute of the design engineer. Therefore schedules and plans built on historic data tailored by the design engineer are asking for trouble. Any schedule analysis must be based on the non-finite probabilities of achieving an objective (i.e., the functionality).

Erroneous Man Month estimates are caused by increasing system complexities and the number of interrelated processes. Stated simply, no amount of effort is going to change the fact that the gestation period for a baby is nine months. As illustrated in Fig. 5.5, as the Functional Complexity increases, and the number of interrelated processes increases, there are some engineering problems that are actually worsened by increasing the level of effort (adding more engineers). Clearly, when designing the engineering process, attention must be given to minimizing the relationships between the individual processes, even if it impacts the product technology [32].

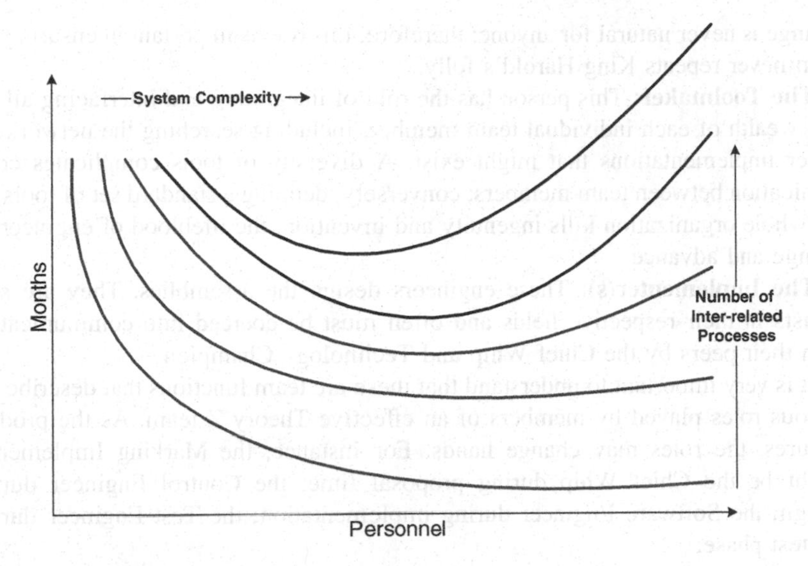

Fig. 5.5 Man-month estimates vary with program complexity

Testing is an undervalued effort. Whatever the time is scheduled for design it must include the design of built-in-tests. In short, Test Driven Design must be a priority of any system. Only for the simplest of systems does a design engineer fully understand what he has made. Testing is an invaluable component of design and, with statistical monitoring, can provide a measure of the quality of the product.

Courage and Common Sense is required to ensure naiveté does not overwhelm engineering estimates.

Based on his experience, Brooks suggests a "Surgical Team"[2] designed to be no larger than ten people operating on one Enterprise Domain within a system. In order to move IPTs into the Theory Z paradigm, the following roles are identified:

The Chief Whip: This role ensures the team focus is never lost, providing the principal interface to program management, interfacing teams, and service organizations such as Quality Assurance. The Chief Whip is **NOT** solely responsible for defining the team objectives as that is a matter of consensus, but it is his responsibility to ensure that there **IS** a consensus. The Chief Whip is also responsible for ensuring conceptual integrity is maintained. All too often, architectures become a patchwork quilt of improvements that hinder maintenance and add little in terms of quality. The QFD process, described later, can help the Chief Whip to control improvement and develop consensus.

The Technology Champion: This role gleans new technology and concepts for the team. He/she must argue for its use and even demonstrate its applicability.

---

[2] The term Surgical Team draws on the hospital analogy of a highly specialized team of professionals, each with a specific skill, that cooperates and collaborates to achieve specific goals.

Change is never natural for anyone; therefore, this role is important in ensuring the team never repeats King Harold's folly.

The Toolmaker: This person has the role of integrating and interfacing all the tool wealth of each individual team member, including searching the network for better implementations that might exist. A diversity of tools complicates communication between team members; conversely, defining a standard set of tools for the whole organization kills ingenuity and invention, thc lifcblood of engineering change and advance.

The Implementer(s): These engineers design the assemblies. They are specialists in their respective fields and often must be coerced into communicating with their peers by the Chief Whip and Technology Champion.

It is very important to understand that these are team functions that describe the various roles played by members of an effective Theory Z team. As the product matures, the roles may change hands. For instance, the Marking Implementer might be the Chief Whip during proposal time; the Control Engineer during design; the Software Engineer during implementation; the Test Engineer during the test phase.

5.2 The Nature of Consensus Engineering

Before embarking on the technical impacts of the Theory Z organization, there are two concerns that require attention:

- The affect that a shift from a command control structure has upon regulation; and
- The problem of defining the organization's mission.

To address these, first we review the consensus problem, and then the solution using the Quality Function Deployment (QFD) process.

5.2.1 What Does It Mean to have a Common Vision

The problem appears at least once a year for the engineering organization; management wants a detailed plan that goes out for 5 years. The trouble is the work, especially research, has not been performed to be able to see beyond 1 or 2 years, and the competition is planning the next 20 years.

This is not uncommon in bureaucracies. Remember, the bureaucracy is designed for invariant systems and business environments. Consequently, vision and engineering don't seem to mix. Clearly, a Theory Z organization requires a clear definition of "vision" that is both practical and futuristic; without it there will be a compromise and disagreement. One such definition is given in Table 5.2.

Table 5.2 Types of Theory Z Vision

Nearsighted vision: objectives	Farsighted vision: objectives
• Expressed in numbers	• Expressed in paradigms
• Practical and mundane	• Idealistic and futuristic
• Requires group consensus	• Requires imagination and time

The nearsighted vision is defined by a set of objectives associated with work capital. As shown in Fig. 5.3, these objectives are the basis for managing work and can be constructed to manage product teams without resorting to bureaucratic methods.

The farsighted vision is the domain of the R&D group, marketing, and the product teams. Whereas objectives are associated with engineering, destinations are associated with Imagineering. Bill Conway [13] describes Imagineering as "a key concept in problem solving and involves the visualization of a process, procedure, or operation with all wasted eliminated" [16]. Another way to imagine[3] is to dream of the extraordinary and impossible that would transform the business and generate new markets. The farsighted vision must be owned by both the R&D group and the product teams; they are not developed top-down as in a bureaucracy, but are owned by the organization, the team, and the individual.

The challenge in today's engineering world is to use the Information Technology available to capture, maintain, and develop a corporation's knowledge capital, and to integrate objectives and destinations. The book will describe methods for this, along with new tools for its implementation.

5.2.2 Examples of Theory Z Vision

The idea of knowledge capital and Imagineering is not new. Here are some examples of organizational methods for their implementation.

Gatekeepers: are the most common visionaries to be found in engineering; they link organizations to the technical world at large. Gatekeepers have been characterized by Tanzik [47] as: high performers recognized within the organizations, tenure in the organization for about five years, have a high status level; and tend to be first line supervisors that maintain an up-to-date understanding of new technologies. It has been noted by [14], however, that without management leadership, much of the organization can remain suspicious and not willing to listen to the gatekeepers. Consequently, the Gatekeeper's role is by its nature informal and underground in organizations where they are not recognized by management.

[3] Although many people believe Walt Disney coined this term, it was actually popularized by Alcoa in the 1940s.

The Skunkworks: instituted within several large companies have yielded significant results. In particular, Johnson [36] led small teams at Lockheed to build several aircraft. In particular: the U-2 employed 50 people; the non-trivial SR-71 program employed only 135 people. Johnson describes the skunkworks philosophy as follows:

> The ability to make immediate decisions and put them into rapid effect is basic to our successful operation. Working with a limited number of especially capable and responsible people is another requirement. Reducing reports and other paperwork to a minimum, and including the entire force into the project, stage by stage, for an overall high morale are other basics. With small groups of good people you can work quickly and keep close control over every aspect of the project.

The Skunkworks philosophy has been tremendously successful, but it has suffered severely over the decades from bureaucratic "help".

5.2.3 Shifting Paradigms for Agile System Design

A paradigm is a set of rules used to understand a system. If we do not understand the system we are working with, it is because the paradigm we are using is dysfunctional. The bureaucracy, for example, is an organizational paradigm as is the IPT method. Similarly, the computer tools we use are designed for a particular paradigm (e.g., linear analysis tools are meant for designing in the frequency domain). Mastering change can be viewed as being proficient in paradigm shifting.

Barker [3] has described paradigms in the following way:

1. *Unsolvable problems* are identified by the prevailing paradigm, which acts as a trigger to search for a new paradigm.
2. *Paradigm Shifters* are usually outsiders who come from the fringes, not the center of the prevailing paradigm community.
3. *Pioneers use intuition* when formulating a new paradigm as there is never enough proof to form a rational decision.

The objectives described in Table 5.2 form the operations paradigm. If there is no rational, measurable basis for those objectives, then it is unlikely that the organization will accept the news that the paradigm is broken. Without that basis, anyone who ventures to suggest that the paradigm is broken will be considered to be making an unwarranted personal attack, as the paradigm is powerful in defining the engineering culture. The consequence is paradigm paralysis, where an organization is blinded by the apparent success of its own culture.

Our Theory Z concepts of **Codeterminism** and **Open Functionbases** form an essential part of the solution to paradigm paralysis. By having a rational basis for objectives and treating captured knowledge as group owned, the various product teams as well as the R&D groups can browse and identify shortcomings in the paradigms and suggest alternatives based on objective facts.

The destinations described in Table 5.2 form the future paradigms. Often there is no rational basis, it is intuitive, and evidence might even be anecdotal in nature. This is why a Theory Z organization has an R&D group that is not dependent upon product areas for approval of their funding. Intuition is anathema to the product manager. The Theory Z concepts of automation are essential here to allow the product teams time to communicate with the R&D group so as not to slip schedules agreed to with the product manager.

5.3 Quality Functional Deployment for Theory Z

Paradigm shifting has become the essence of the successful organization. It is paradigm shifting that enables an organization to "change with the times", staying competitive and adapting to new business environments. The important question is how we identify broken paradigms and forge the organization's vision at the same time. The method proposed here is to use QFD methods.

5.3.1 The Role of QFD in Theory Z

QFD consists of seven tools, as described by King [38], that can be used to address various aspects of the design process. They are system analysis tools that identify the objectives to a fine degree of detail and consist of:

1. Affinity diagrams for brainstorming.
2. Interrelationship digraphs to ensure the problem is understood.
3. Tree diagrams to hierarchically structure ideas and system components.
4. Matrix diagrams to correlate the trees.
5. Matrix data analysis.
6. Decision program charts to capture failure modes.
7. The arrow diagram to map out the final process.

For a bureaucratic organization, QFD may have limited value and be considered too expensive to implement, hence its lack of acceptance within engineering organizations. But for the Theory Z organization, that has to manage change, QFD provides a rational basis for continual improvement. It provides a discipline for integrating the voice of the customer with the voice of the engineer. It enables the team to understand what the customer really wants. It also enables the team to evaluate and integrate one another's values to form a solid consensus, thereby integrating vision and constructively addressing the culture clash problem.

QFD also provides a rational basis for identifying new technology and concepts. The product teams can work in concert with the R&D group to identify the ideas with most promise, thereby pre-empting the "not invented" here syndrome.

If properly utilized as a systems tool, QFD provides a priori traceability from the customer's demands to the final product before design and manufacturing have started. A corollary to this is that QFD facilitates reuse when implemented using an automated, hypertext-driven system. Any new customer demand can immediately be reviewed relative to existing designs and the closest fit found.

Another element of QFD is the management of objectives such that they define the desired robustness characteristics. Robustness is a key feature in Concurrent Engineering. It pertains to the resilience of the engineering processes and product to uncertainties and unexpected changes.

Therefore, the *QFD* methods, and the *Functionbases* they generate, are the principal tools for integrating the organizational elements described in Fig. 5.3.

Figure 5.6 illustrates the balance gained by using QFD as the goals of Concurrent Engineering, which can be defined as "the systematic approach to the integrated concurrent design of products and their related processes, including manufacturing and support" [20–22].

The QFD process can be used by any organization, but it promises the best return on investment when used by an organization that is optimized for change.

First, consistent with Likert's System 4, QFD facilitates communications vertically and horizontally, especially when it is accessible over a network.

Second, consistent with Theory Z, the QFD method should facilitate extended communication such that a consensus is developed.

This is why the load-bearing beam in Fig. 5.6 is titled "Cooperation and Discipline" and the operative term is "should facilitate". Such a return on investment is unlikely in a Theory X or Theory Y organization.

5.3.2 *Cost Comparison*

The use of the QFD method and Concurrent Engineering is really a move away from the sequential design approach to an incremental design approach (which many might call an "agile" approach) [12].

The sequential method, often called stove piping, is used by organizations composed of tight functional groups. The product design is passed from function to

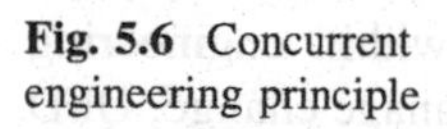
Fig. 5.6 Concurrent engineering principle

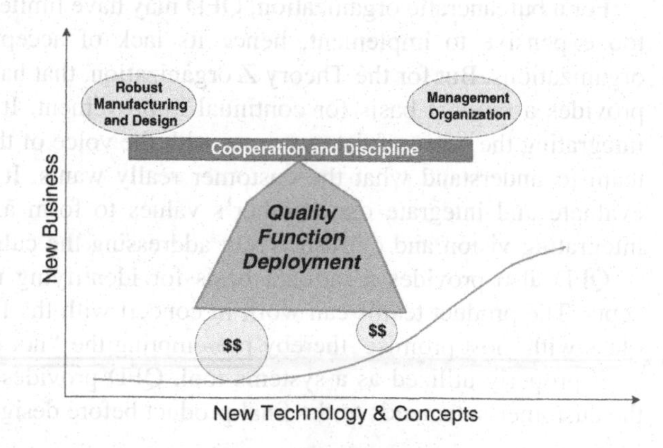

function, each adding a layer of detail. If adding a layer uncovers an error in one of the earlier layers, the whole design has to be peeled back, corrected, and then re-layered.

The incremental approach uses the multifunctional (product) team to pre-empt design faults, especially when the QFD process is utilized.

The overall effect is to change the normalized cost profiles, as illustrated in Fig. 5.7. Sequential design (Fig. 5.7a) is alluring in that the initial costs are low, giving the program manager a false sense of security. Even if cost and schedule are monitored and on track the schedule never includes the hidden dangers of sequential design. As the design matures, more problems surface causing cost to grow exponentially.

Incremental design changes the nature of the cost-incurred curve (Fig. 5.7b). The method is more expensive initially, leaving the program manager feeling insecure. The program manager's insecurity is heightened by the fact that the incremental design still matures slowly; therefore, less work appears to be completed early in the program. But unlike the sequential method, the costs stabilize (i.e., there are few hidden flaws left to be discovered).

Another problem with sequential design is that it commits costs early in the program. With incremental design, more time is spent on the initial design; therefore, costs are not committed unto the team is certain about the requirements and their correctness.

During the initial phases, incremental design requires commitment on the part of the organization and faith on the part of the engineering team, before payoff is seen in the later phases.

5.3.3 *QFD Matrix Method Outline*

The purpose of describing the matrices is not to give a detailed explanation of the QFD process or use. That can be found in King's work. The purpose is to show how the QFD matrices provide the information that binds together the organizational team.

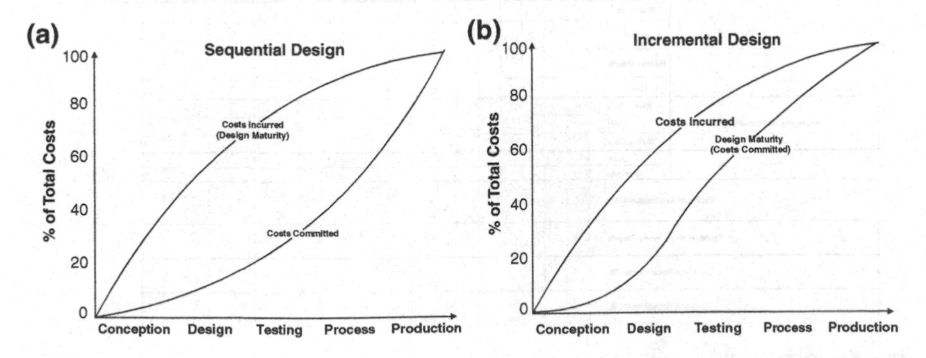

Fig. 5.7 **a** Cost profiles utilizing sequential engineering. **b** Changing cost profiles through concurrent engineering

In King's system of matrices, the A1 matrix (often called the House of Quality matrix), which we illustrate in Fig. 5.8, correlates the Customer's Demands to the Requirements [including Quality Attributes, (i.e., non-functional requirements)]. The matrix is first completed horizontally to determine the most important customer demands and their relative weighting. Then the matrix is completed vertically to determine the most important requirements and their relative importance.

This matrix is completed with the concurrence of the customer, program management, systems engineering, and the product teams responsible for the engineering. The important attributes of the Quality Matrix are described as follows:

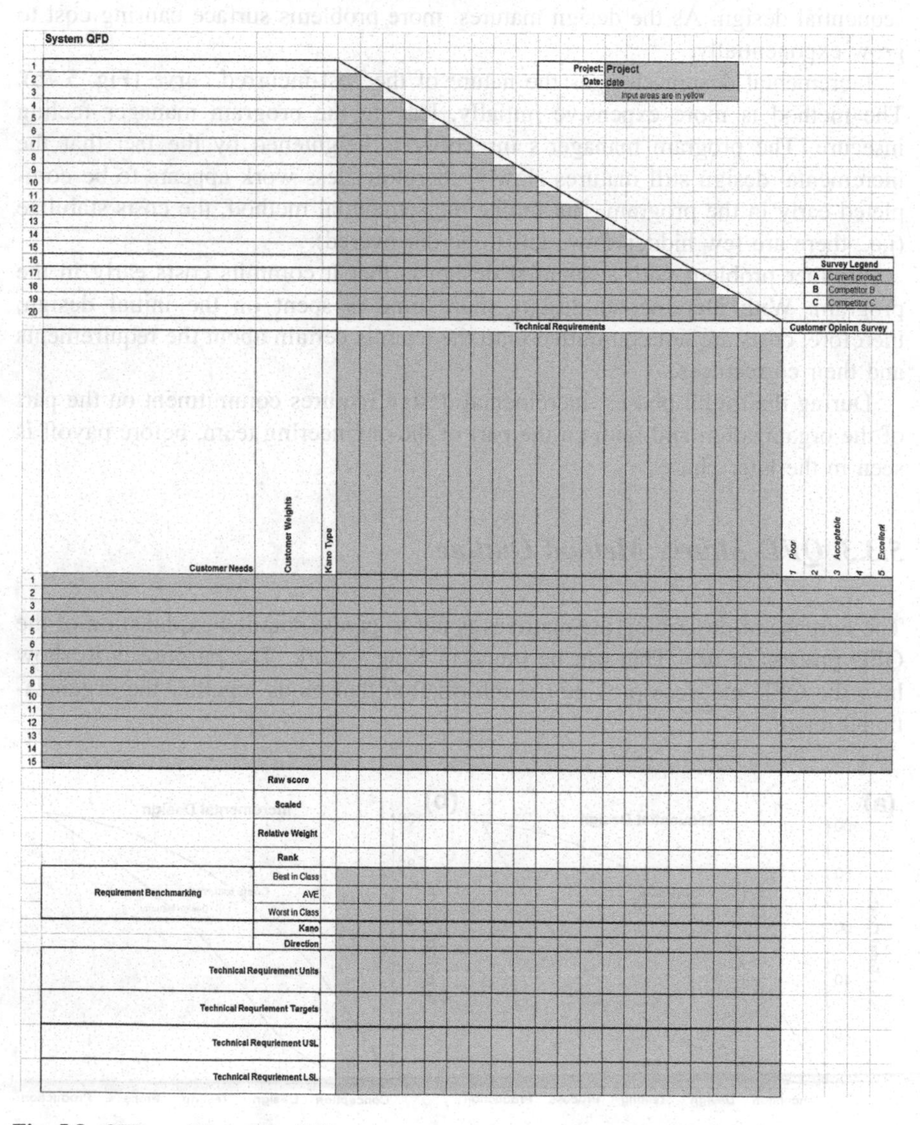

Fig. 5.8 QFD quality matrix (A1)

The Quality Attributes act as robustness drivers. They can come from both customer quality requirements and from statistical analysis to create robustness in the system design, such that the Enterprise Domains are not unraveled by dispersions in the performance of the interfacing Enterprise Domains. Invariably, it is the Quality Attributes that cause the greatest disruption to engineering and drive the overall design of the system and Enterprise Domains. As the program matures, the customer and designer discover the need for better, higher performance. As the customer demands and functions change and adapt, the software and hardware components of the system have to change and adapt to accommodate the new quality attributes and changing functional requirements.

The Quality Matrix provides a roadmap for Continual Improvement. Current design discipline dictates a pragmatic approach to engineering. Therefore, requirements are seen only as absolutes, either "they are met" or "they are not met." The approach described above is focused on "how well" we meet the requirements thereby providing information on what requires improvement and how new technology might improve the quality.

The A2 Matrix is used as a crosscheck, ensuring all the Quality Attributes and Enterprise Domain Functions have been documented and correlated. The A2 Matrix is illustrated in Fig. 5.9.

The A3 Matrix (Fig. 5.10) is used to cross-correlate the Quality Attributes. In this manner, a notification map can be created, enabling the design engineers to manage the interrelationships between Quality Attributes.

The first three matrices are important for product (hardware or service) management. The A4 matrix develops the second level of detail, identifying the hardware and software components required for Enterprise component design and implementation. An example is given in Fig. 5.11.

Other matrices can be developed to evaluate new concepts and new technologies. Each of the three trees, Quality Attributes, Functionality, and Enterprise Components, are evaluated independently against new ideas. The process

Fig. 5.9 Functions and quality attributes matrix (A8)

		Quality Characteristics						
		Reliability	Maintainability	Flexibility	Net Centricity	Scalability	Integrity	Availability
Functions	Mission Management							
	Navigation							
	Command & Control							
	Security Services							
	Data Services							
	Web Services							

Fig. 5.10 The quality attribute cross-correlation matrix (A3)

	Reliability	Maintainability	Flexibility	Net Centricity	Scalability	Integrity	Availability
Reliability	X						X
Maintainability		X					
Flexibility			X		X		
Net Centricity			X	X			
Scalability			X		X		
Integrity	X					X	
Availability							X

facilitates a rational evaluation to help the team members overcome their prejudices, forging the farsighted vision required to define destinations for the organization.

As will be seen, the QFD matrices form the basis for quality design and an information database.

5.4 Summary: Organizational Transition

The engineering organization's ability to change rests more than ever on the education of its engineers and managers, meaning "to provide them with knowledge and training". Education will differentiate the learning organizations that implement Theory Z from older, more bureaucratic organizations. The process of becoming a learning organization will include the following:

Implement QFD to capture the knowledge of experienced engineers to form a computerized Functionbase. Such an implementation will increase the productivity of all the members of the organization as information becomes more readily available.

Automate Engineering Processes to free the engineers from mundane tasks. Historically, automation has been viewed as a threat to one's livelihood. However, automation has always resulted in the user moving on to loftier endeavors. In the case of the engineer, the organizational goal should be to enable the engineer to spend at least 30 % of his/her working hours on education.

Develop autonomous product teams whose budget is not geared to time and motion studies, but to the perceived value of their objectives (as determined through the QFD process). The team must be free to generate their work plans and resource management plans. This freedom must include management of their tools and training requirements.

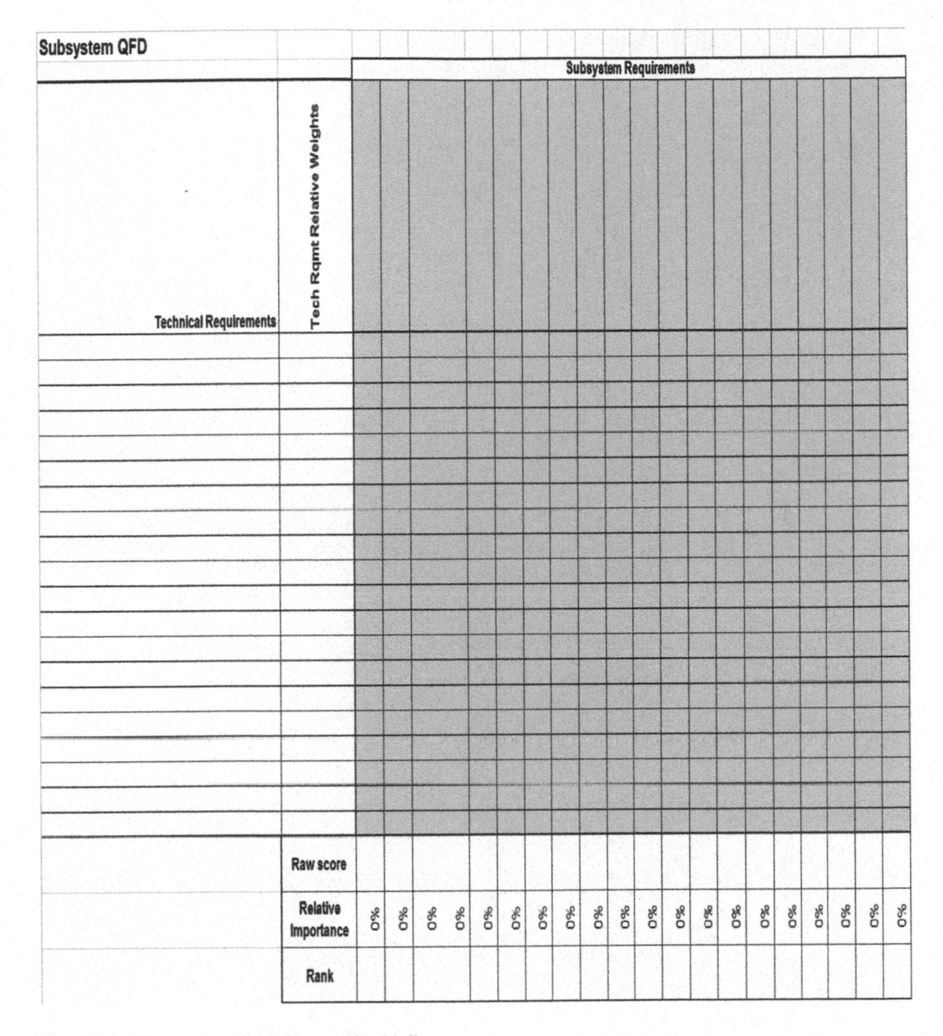

Fig. 5.11 Enterprise domain matrix (A4)

Develop Information Technology to provide a basis for cooperation and discipline across the organization. The Theory Z organization might be viewed as a computer-based matrix of product teams and critical skill departments. The information technology is not just used for automation, but also for building a Functionbase that is available to all.

Promote Continual Process and Product Improvement with the customer and within the organization where the key parameter is the Quality Characteristic. This is especially valid for the product team for whom the quality characteristic is also the team objective. This provides focus for teams' education and cooperation with the R&D groups.

Chapter 6
Modern Design Methodologies: Information and Knowledge Management

Where is the life we have lost in living?
Where is the wisdom we have lost in knowledge?
Where is the knowledge we have lost in information.
T.S. Eliot, "The Rock" (1934).

6.1 The Knowledge Paradigm

6.1.1 The Information Jam

Knowledge is an orderly synthesis of information; it is an abstraction of our understanding gained through experience and study. Information is an orderly collection of facts that form databases (whether in a computer or in our heads); it provides input to our processes. Information Technology is the means used to gather and manipulate data to form knowledge [20]. Wisdom depends on patient osmosis.

Information about the cannon existed in Europe for many years before it was put to effective use. Information about Roman–Greek philosophy existed before the Renaissance. In each case, improved communication was the catalyst for change. The Renaissance catalyst was trade routes in Italy between Europe and the East. This is illustrated in Fig. 6.1 in hierarchical process–product terms.

During the industrial revolution, Burns [10] further invigorated information generation through the use of steam power. The steam powered newspaper arrived in 1814 with the first copy of The Times in London. Since then, numerous technologies have accelerated information generation to the point that we are becoming overwhelmed. It has become too easy to copy data and forward it to colleagues. Did they need it? Is it a waste? Remember waste identifies low quality. In addition, low quality requires rework and therefore, lost opportunities to improve due to the time invested to rework.

However, the proliferation of information has enabled the development of expertise. It is possible to find data on almost any subject across the Web. Figure 6.2 illustrates the change as computers, information systems, and the World-Wide-Web have displaced books as the basis for information; administrative assistants have displaced laborers as the predominant component in society; and managers have displaced company owners as owners of the product and holding the "power."

J. A. Crowder and S. Friess, *Systems Engineering Agile Design Methodologies*,
DOI: 10.1007/978-1-4614-6663-5_6, © Springer Science+Business Media New York 2013

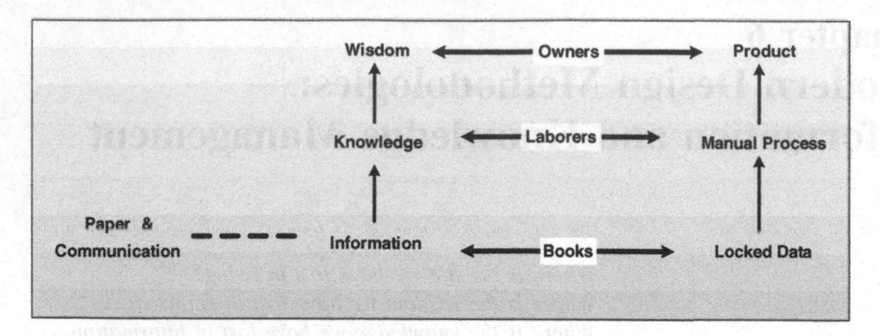

Fig. 6.1 Pre-industrial business paradigm

6.1.2 The Horizontal Integration of Knowledge

Perhaps the simplest example of how power is being shifted by computer technology is to consider communication rights. The telephone (including cell phones) and email are indispensable components in getting work done. This becomes painfully evident in third world countries where telephones, cell towers, and Internet connectivity are scarce; there are no beepers, no voice mail, and no email [25].

In modern companies workers have communications rights, including long distance and international communication through email and web-based connectivity without asking permission. The engineer can now meet colleagues electronically and develop relationships without ever meeting them face to face (although they can meet virtually face to face through video conferencing).

This newfound freedom in communications has provided a plethora of opportunities. Web sites can be found that are full of any information you might be looking for. News events are published onto the Web as they are happening. Streaming video is available across the world on any event.

Suddenly we are living in an electronic village. The shy person is not afraid to speak up, as they are secure in their electronic village. Also, distance no longer prevents casual conversation worldwide as you can simply open up your virtual cottage window and simply chat with the electronic passerby. In addition, one can

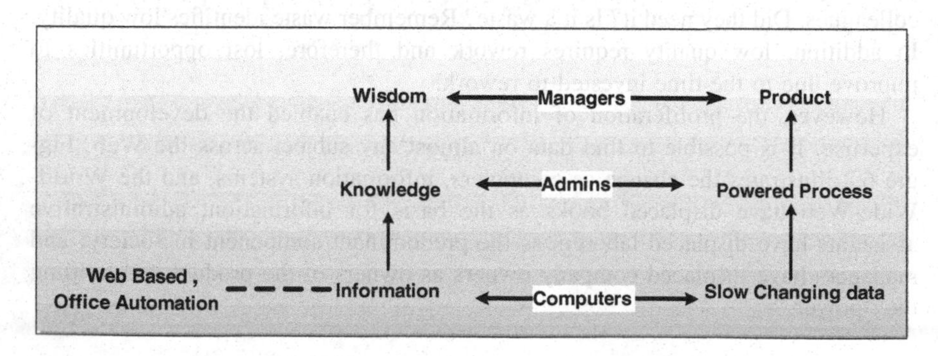

Fig. 6.2 The post-industrial business paradigm

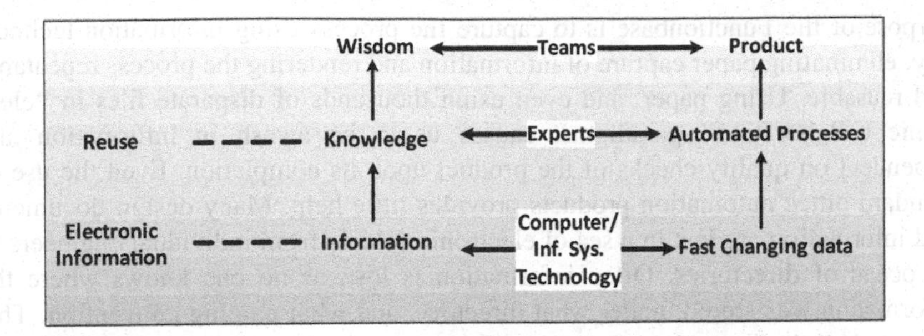

Fig. 6.3 Business Paradigm for the information age

browse electronic databases from anywhere in the world like walking a trail through the local park.

Current computer technology enables us to redeem the time we need to discover what's out there. It provides a broad horizontal integration of the workplace and the rest of life.

Within an organization, the horizontal integration is more focused. Just as the manager replaced the owner at the turn of the century, today the team is replacing the manager. Teams can be formed on the fly as engineers and scientists find one another over the network, identify their common vision, and then they can disband once they reach their objective.

The bureaucracy, that integrated the clerks, is being replaced by the electronic network of experts [22] and systems engineers. The middle manager will have a new role on the team: facilitating the QFD sessions and keeping relative focus for the team. This is illustrated in Fig. 6.3, where teams replace the manager and experts replace the clerk.

However, there is another important distinction. The Industrial Revolution enabled the rapid proliferation of information, and now the Information revolution enables the rapid reuse of information and knowledge.

There is need for Functionbases, which captures rules for using information and knowledge. In this manner a high level of organization can be imposed upon the information for rapid reuse that provides history, background, and context for the information. Automation then enables members of teams to use one another's functions with ease. Less time is wasted on the detail of how; more time is spent on the details of why [21]. This is how teams self-manage, replacing managers as owners of the product.

6.2 Engineering Tools for Functionbases

6.2.1 A Simple Functionbase Approach

The paper paradigm pervades every aspect of our lives. Efficiency improvements often lead to the generation of more paper as quality improvements are sought. The purpose of the Functionbase is to capture the process using information technology, eliminating paper capture of information and rendering the process repeatable and reusable. Using paper, and even using thousands of disparate files in "electronic collaboration" paradigms, causes us to be awash in information and dependent on quality checks of the product upon its completion. Even the use of standard office automation products provides little help. Many design documents and information are lost in a sea of electronic folders from individual engineers in an ocean of directories. Often information is lost, as no one knows where the information was stored, under what directory, and what naming convention. This process is little better than paper reports in separate filing cabinets.

A simple example is provided by the spreadsheet. Early spreadsheets were databases that helped in the layout and basic organization of information. One early innovation was to add Macros that could repeat specific user operations. All that was required was training on the computer application the particular operations, and it could then be repeated on demand. Voila, automation. Ensuring the Macro was correct is putting quality into the Functionbase. The product inherited the quality and did not require checking. Checking had become a waste, quality had been improved.

The automation for engineering functions and disciplines is similar. For functions, such as Mission Management or Command and Control, build a database of functionality and append a Functionbase, such that the design is automated. What's more, the process tool must be self-documenting. If the engineer has to "copy" information into a flat-file report, then there is opportunity for error and quality control checking will have to be reinstated. The team need only check the original Functionbase knowing the new data propagation will be correct.

This method has been tested using MATLAB® and METADESIGN® for the design of a Metrics Analysis Systems. These two tools easily integrate together to form a Functionbase that captures the Metrics analysis process cradle to grave. The following attributes have been noted [18]:

(1) *Productivity* was increased up to tenfold.
(2) *Repeatability* enabled inexperienced engineers and team members (including the customer) to reliably reuse the design.
(3) *Shareability* has enabled the Functionbase to be the contract deliverable.
(4) *Verification and Validation* engineers could now review the Functionbase for design methodologies, codes, results, etc., a complete self-contained package.

6.2.2 Engineering Tools: The Engineer's Sandbox

Engineering tools like MATLAB® and IDL® serve two important functions. They can be used as executable specification to software engineers and provides the design engineer with a sandbox for experiments as illustrated in Fig. 6.4.

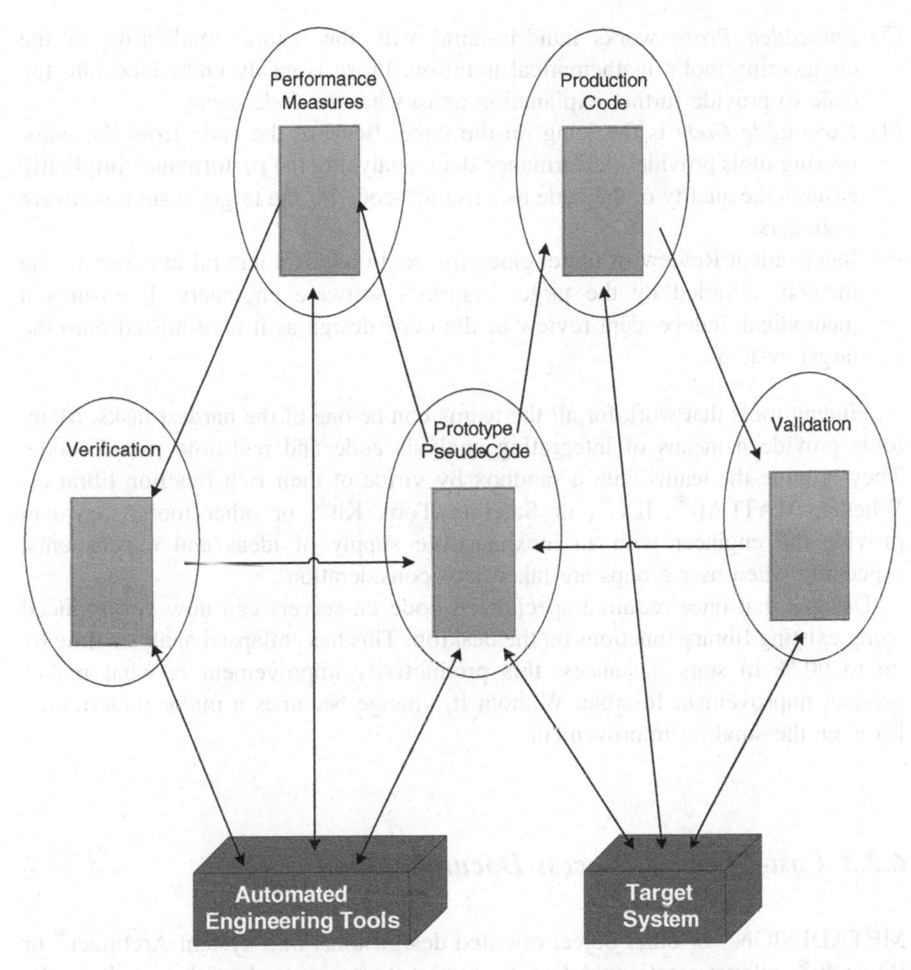

Fig. 6.4 Integrated analysis and design methodology using automated engineering tools

The use of engineering tools as executable specifications has many advantages. Code and results produced from such tools can be used to validate unified modeling language (UML) flow diagrams, utilizing the engineering tool code as pseudo-code to validate the UML design documentation for the subsystem. The advantages are:

(1) *Single Representation* of the real-time code requirements. The Functionbase from the engineering tools becomes pseudo-code to the target system software engineers and an executable analysis code to the systems design engineer. There is no redundant data to confuse the team when it comes to resolving design ambiguities.
(2) *Mathematical Notation* makes it easy to correlate the design code with the mathematical derivations. This facilitates verification of requirements.

(3) *Embedded Prose* works hand-in-hand with the natural readability of the engineering tool's mathematical notation. Prose is easily embedded into the code to provide further explanation as to what the code means.
(4) *Executable Code* is the icing on the cake. Because the code from the engineering tools provides performance data, analyzing the performance implicitly ensures the quality of the code as a pseudo-code for the target system software engineers.
(5) Independent Review of the engineering tool code is a natural attribute of the method, afforded by the target system's software engineers. It ensures a methodical, independent review of the code design as it is re-hosted onto the target system.

Finding tools that work for all the teams can be one of the hardest tasks. Many tools provide a means of integrating analysis code and real-time pseudo-code. They provide the teams with a sandbox by virtue of their rich function libraries. Whether MATLAB®, IDL®, or Satellite Took Kit®, or other tools, they can provide the engineer with an inexhaustible supply of ideas and experiments, especially when user groups are taken into consideration.

Designs that once required specialized code on servers can now be produced using existing library functions on the desktop. This has collapsed analysis time by up to 90 % in some instances; this productivity improvement is what makes product improvement feasible. Without it, change becomes a major undertaking for even the smallest improvement.

6.2.3 Cost-Effective Process Documentation

METADESIGN® or other object-oriented design tools like System Architect® or Rapsody®, allows creation of Functionbases that manage knowledge about the design throughout the systems life cycle. However, the components of the Functionbase are best understood by reviewing how it should be used in conjunction with the other tools. Recurring Design is the most common use. This occurs when a new configuration requires a change (say to a Mission Management system).

Figure 6.5 illustrates the process utilizing the Functionbases. Starting at the top of the *Recurring Design* tree, this first page that appears would be instructions on what scripts to run in what order for the engineering tools associated with the Mission Management models. This is accomplished with hyperlinks on the first page to other pages within the Functionbase. For example, the first link should be to the *Configuration Data* page. Entered onto this page is all of the data required to complete the design. The next link should be to the engineering tool script created for the design model and etc., until the design is complete [18]. During the design sequence for the changes, the engineering tools will generate text containing the new set of data for inclusion into the Functionbase for the new design (the changes

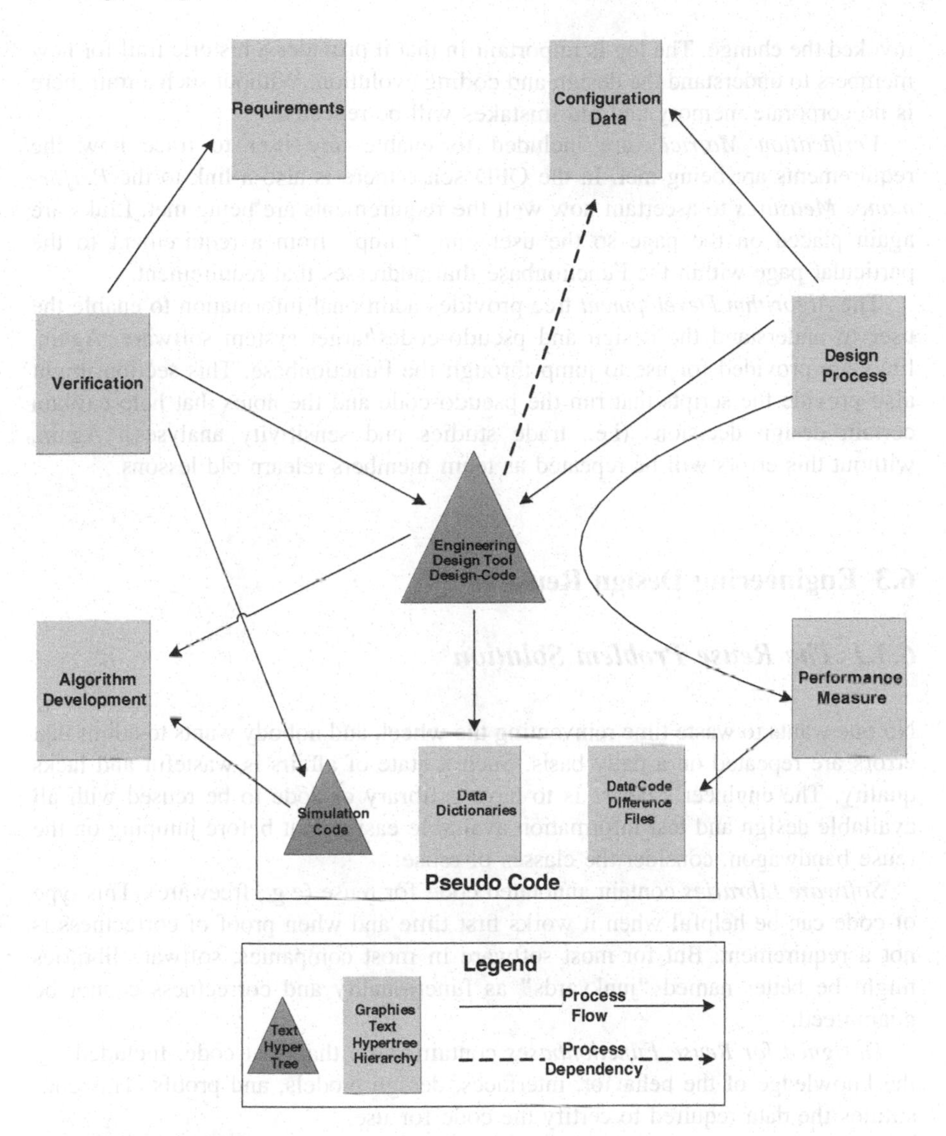

Fig. 6.5 Generalized functionbase reuse

being made). When run, the new scripts that generate the *Performance Measure Data* will use the *Data Dictionary Tree* and the *Pseudo-code Simulation*. In this manner the files used by the target system software engineers are automatically checked through the performance analysis.

On completing the *Recurring Design* exercise, the target system software engineers can access the pseudo-code trees and export the necessary data files. Included are *Difference Files* that show exactly what pieces of data have changed. This includes a change log with references back to the design discrepancies that invoked the change. The log is important in that it provides a historic trail for new members to understand the design and coding evolution. Without such a trail there is no corporate memory and old mistakes will be repeated.

Verification Matrices are included to enable any user to trace how the requirements are being met. In the QFD sense there is also a link to the *Performance Measures* to ascertain how well the requirements are being met. Links are again placed on the page so the user can "jump" from a requirement to the particular page within the Functionbase that addresses that requirement.

The *Algorithm Development* tree provides additional information to enable the user to understand the design and pseudo-codes/target system software. Again, links are provided for use to jump through the Functionbase. This section might also provide the scripts that run the pseudo-code and the notes that help explain certain design decisions (i.e., trade studies and sensitivity analyses). Again, without this errors will be repeated as team members relearn old lessons.

6.3 Engineering Design Reuse

6.3.1 The Reuse Problem Solution

No one wants to waste time reinventing the wheel, and nobody wants to admit that errors are repeated on a daily basis. Such a state of affairs is wasteful and lacks quality. The engineer's desire is to have a library of code to be reused with all available design and test information available easily. But before jumping on the reuse bandwagon, consider the classes of reuse.

Software Libraries contain annotated code for reuse (e.g., freeware). This type of code can be helpful when it works first time and when proof of correctness is not a requirement. But for most software in most companies, software libraries might be better named "junkyards" as functionality and correctness cannot be guaranteed.

Designed for Reuse Functionbases contain more than just code. Included are the knowledge of the behavior, interfaces, design models, and proofs. This constitutes the data required to certify the code for use.

Domain-Specific Solutions are complete hardware–software solutions. The home PC delivered with preloaded operating system and office automation software is an example.

Discussed here will be the second class: Designed for Reuse Functionbases. For this class, the reuse requirements must be carefully defined.

First, knowledge must be defined and included in terms of a mathematical basis. This is why the reuse junkyard is so hard to deal with, as the mathematical basis is not known; therefore, its behavior within a system is unknown. The need for a mathematical basis is the reason engineering tools, such as MATLAB, as an executable functional specification are useful. The engineering tool pseudo-code syntax is mathematically based, and provides insight into the mathematically basis for the design.

Second, interfaces have to be completely defined to avoid errors that involve data flow, priorities, and timing errors at both the highest and lowest levels of the system [28]. Common experience shows interface problems to take over 75–90 % of the errors found after implantation of reuse code with conventional techniques. This drives up cost due to all the investments in manufacturing interface control documents (ICDs). Robust error checking is an essential design-for-reuse feature, as it precludes this type of expensive error. Code correctness must be assured before compiling.

Third, design proofs are required to preclude system errors. A prior verification of code precludes expensive system bugs, such as instabilities. The design proof leverages off the mathematical basis; algorithm correctness must also be assured before compilation.

Four basic tools are required to provide the attributes for a real design-for-reuse paradigm:

(1) A distributed hypermedia tool for workstations to allow creation and tracking of Functionbases.
(2) Engineering design tools like MATLAB, IDL, and Satellite Toolkit for building pseudo-code simulations and prototypes.
(3) Mathematical Equation tool, like Macsyma® for building and encoding mathematical models.
(4) A Case tool used to implement the robust code building routines, needed to preclude errors at their source.

The relations between these tools are illustrated in Fig. 6.6. The reuse methodology integrates the behavior of the two ends of the system. The engineering tools form a mathematical basis for capturing the system's algorithmic behavior. The case tools have a mathematical basis for capturing the target system's behavior. Experiments can quickly be run in the sandbox before time is spent completing the design. The Data Dictionary can be built that complements the engineering tool's functions, and the target code re-ingested into the engineering tool (in actual code C++, JAVA, etc.) for verification. Tests can then be devised in the sandbox for running on the target system for final verification and validation.

6.3.2 Provably Correct Code

The engineering tool's mathematical basis makes it an ideal candidate for experimental design. Part of the reason for this is that there are generally no typing constraints. Conversely, code for target systems require strong typing, hence the Data Dictionary (Fig. 6.5). The experimental nature of the engineering tools is also used at the other end of the life cycle in analyzing data from the target system (see Fig. 6.6).

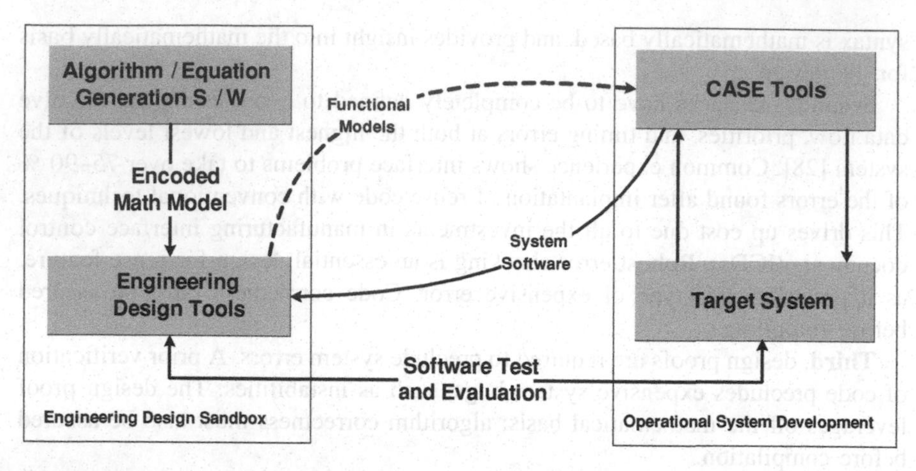

Fig. 6.6 Design for reuse methodology

In between these two life cycle ends is the Case Tool to bridge the gap between the functional architecture and the resource architecture of the target system.

The functional architecture is designed by function hierarchies (called FMmaps) used by the engineering tools and type hierarchies (called TMaps) implied by the engineering tools and used by the Case Tools. Three primitive control structures are used. There is one for defining dependent relationships, one for defining independent relationships, and one for defining decision-making relationships. A formal set of rules associated with these is used to remove design errors from the maps. Because the primitive structures are reliable and because the building mechanisms have formal proofs of correctness, the final system is reliable. Furthermore, all modal viewpoints can be obtained from the FMaps and TMaps (e.g., data flows, control flows, state transitions, etc.) to aid the designer in visualizing the design.

The engineering tool's reuse value is in its mathematical basis. The reuse value of the Case tools is founded in its separation of functional and resource architectures. Once the functional architecture is defined, it can be used on any target system, whose resource architectures are fully reusable by the engineer, when they match the engineering tools functions. Put another way, reuse requires two Functionbases: the engineering tools functions and the case tools functional architectures that capture the system behavior. These two Functionbases capture all the information required to complete specificity of the system, no more, no less. All that is required is the case tool that integrates and analyzes these architectures. This is illustrated in Fig. 6.7.

The analyzer is used during the definition of the TMaps and FMaps to test for consistency and logical correctness before placing the maps in the library. Templates of the particular target system are built and populate the resource architecture library. The Resource Allocation integrates templates from the library and then automatically generates the required source code. Run-time performance

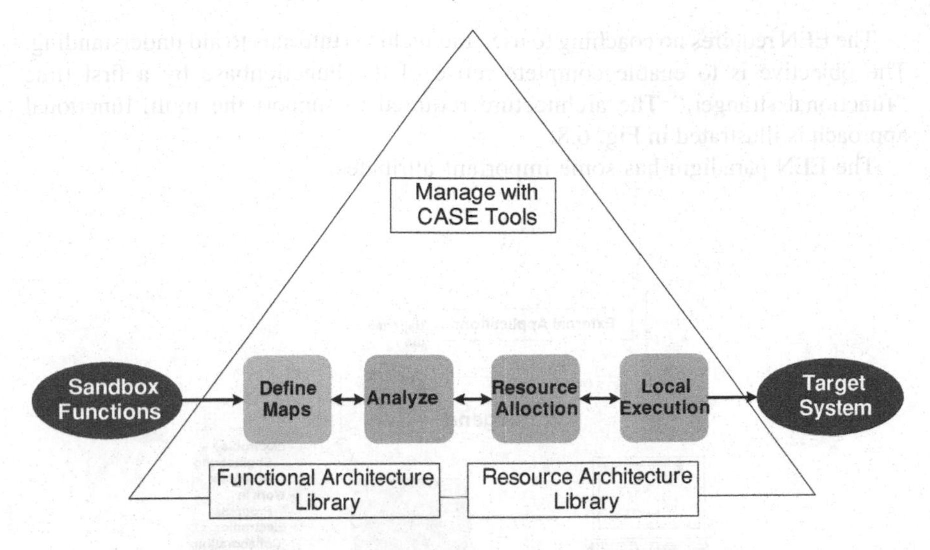

Fig. 6.7 Bridging the sandbox and the target system

analysis of the code can be verified locally to ensure it meets the constraints of the target system (e.g., timing).

A synergism is realized by using the engineering tools and the case tools together. They provide a design process with built-in quality. When combined with the QFD methodologies, this paradigm provides a framework for automation, aiding the organization in defining and implementing process improvements [37, 46].

6.4 Groupware Utilizing Knowledge Management

6.4.1 The Electronic Engineering Notebook

The Functionbase is structured to look like a technical memorandum at its top level. This enables anyone in the team to reuse the Functionbase as an electronic memo, and play what-if games to better understand the design. It also enables rapid-generation of the design, because the process is built into the Functionbases. Careful management ensures quality with the process (i.e., quality is built into the Functionbase); therefore, time is not wasted checking the pseudo-code files or checking the performance analysis.

The corollary for all this is the notion of a hypermedia, electronic engineering notebook. This electronic engineering notebook (EEN) includes the software tools and codes, such that an engineering activity performed using the EEN can be repeated without additional knowledge. It contains the information about what it is, what it means, and how to use it. This Knowledge Management System provides the hypermedia capability to provide the functionality described above.

The EEN requires no coaching to use, and includes tutorials to aid understanding. The objective is to enable complete reuse of the Functionbase by a first time "functional stranger." The architecture required to support the multi-functional approach is illustrated in Fig. 6.8.

The EEN paradigm has some important attributes:

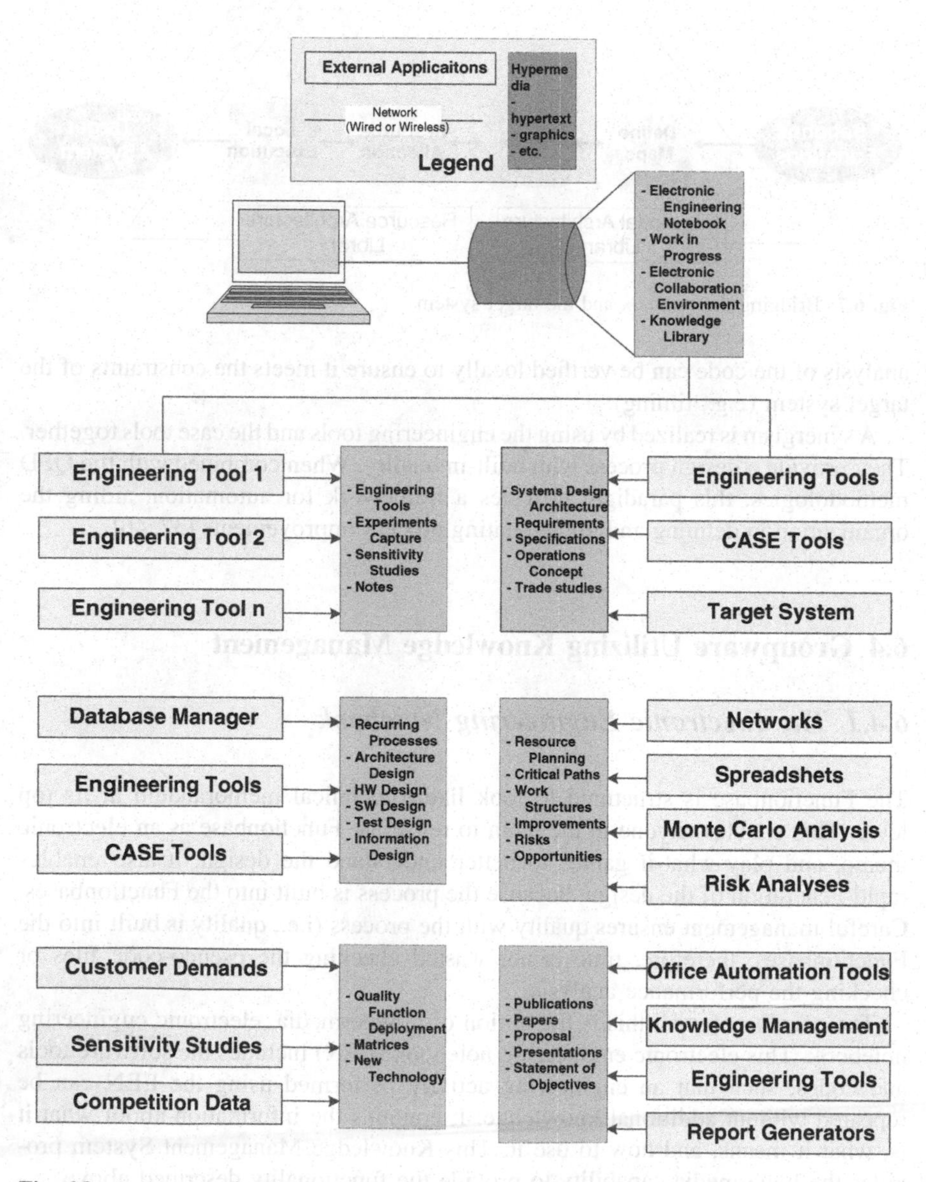

Fig. 6.8 Architecture for the electronic engineer's notebook

Process Objects are directories containing all the files and directories required to complete a discrete design or analysis. The object is a collection of linked text, graphics, and applications.

Transparent Network Objects enable team members to browse one another's Process Objects. Virtual documents can be built and printed if required, using links that traverse the network to integrate Process Objects.

Process Automation is provided using a scripting language. The EEN can be taught the process such that designs are automated.

Functionbase Security is provided to ensure data integrity is not violated.

Anyone who understands the above should see that this is infinitely doable utilizing today's Web Services and Java Scripting, coupled with Office Automation Tools that include hyperlinking capabilities.

6.4.2 Required Engineering and Knowledge Management Functions

To support a team, the EEN must contain a broad range of user functions. Two classifications can be made: engineering functions that support the program and knowledge management functions that support the engineer. These are, of course, correlated, as illustrated in Fig. 6.9.

The major categories for Required Engineering, shown in Fig. 6.9 are:

Interactive Experiments refers to the process whereby an engineer can create a "living" notebook in an online environment to retain analytical results and interleave comments and observations. This serves as a replacement of the classic Engineer's Notebook. This is accomplished in a real-time environment while analyses are actively being performed. This also allows creation of hypertext links to any other relevant material to provide a dynamically growing structure of cross-linked reference material. Analysis differs from the sandbox only in terms of rigor.

Recurring Engineering refers to engineering tasks such as performance analyses, design analyses, and the measurement of performance for quality assurance, all of which can be automated.

Non-recurring Engineering—there are three domains. First, the Functionbase that contains the QFD matrices governing the integration of the organizational elements and the incorporation of new technologies. Second, resource planning requires worksheets, networks, and analyses essential to the determination of the work plans, the integration and management of resources, and the incorporation of improvements. Third, the design phase which are all those activities performed by the engineer, such as building a Functional Architecture Library.

Publications are the release of engineering requiring navigation paths to select subsets of information. This can be performed in the knowledge management domain (Web Services) or can involve Case Tools and may be published in electronic media form.

The major categories for Knowledge Management Functions, shown in Fig. 6.9 are:

Data Management involves integration of both graphics and text into a hypermedia Functionbase. Data access must be capable of being automated.

Function Management involves the development and structuring of procedures. An example is code management (Configuration Management), where code should be captured within a tree and executed directly (i.e., through a button on the screen).

Linked Data Structures are methods of data organization, which consistently allow the monitoring and modification of data interrelationships and interdependencies. The structures are composed of trees that form objects and navigation paths across the trees.

		Knowledge Management Functions								
	Required Engineering:	Data Management	Function Management	Linked Data Structures	Unified File Systems	Email/Web Services	Auto Notification	Quality Assurance	CASE Tool Management	Engineering Tool Interfaces
Interactive Experiments	Sandbox	X	X	X					X	X
	Analysis	X	X	X	X	X			X	X
Recuring Engineering	Automated Processes	X	X	X	X		X	X	X	X
	System Data Coniguration	X		X	X		X	X	X	X
Non-Recurring Engineering	QFD	X		X	X	X		X	X	X
	Resource Management	X				X	X		X	X
	S/W Designs	X	X	X	X	X		X	X	X
Publications	Design Release	X		X	X		X			
	Presentations			X	X	X	X		X	X
	Papers/Memos/Directives	X	X	X		X	X		X	X

Fig. 6.9 Electronic engineering notebook functionality

Single File System means a data item exists in one place only. There are no electronic copies (except for backup), as this would compromise the Functionbase. All processes refer to the one copy.

Automatic Notification is the ability to automatically notify a designated user, or list of users, when a particular, selectable, event has occurred. This is used when a browser leaves notes and comments, or when data someone else is dependent on is changed by the originator.

Quality Assurance ensures engineering is not released until all Quality Characteristics have been verified. Because the QFD established the minimum performance requirements, all that is required is the knowledge management system within the EEN to check that each Process Capability Index (Cp) parameter is greater than one (Cp will be explained later, see Fig. 7.2). Anything less would indicate nonconformance.

Tool Interface is the ability to invoke Case Tool or Engineering Tool programs by spawning a separate process. This includes the ability to interface with the program by sending input to and receiving input from that program, from within the EEN. The data so transmitted (and stored in the EEN) must be both alphanumeric and graphical. This program should be capable of being run in an interactive mode as well as batch.

6.4.3 Groupware Lessons Using Knowledge Management

The Need to use Metaphors: As with any paradigm shift, Knowledge Management has been received with skepticism. The perception that the Knowledge Management paradigm is an improvement is more readily perceived when it is packaged within a familiar metaphor, hence the Electronic Engineer's Notebook name. Once engineers begin to experience it, and management sees it, the community will accept it as a productivity booster.

As a Life Cycle Tool: Knowledge Management will gain acceptance across the industry, but not without some pressure being applied to the engineers that use it. This can be attributed to the general reluctance to change we exhibit as humans. Making the Knowledge Management paradigm broad in application will help in that an engineer will be able to use it for a task that is personally comfortable (e.g., making viewgraphs). Once on the learning curve, the engineer will grasp some of the other more subtle aspects of the paradigm with growing experience.

Advancing Communications: Being able to pass reusable knowledge to a peer will have advantages on both the Intranet and Internet. Whole trees are treated as objects, and can be shipped out to teammates in organizations in any geographically diverse location, or are used in conjunction with a Virtual Development Environment. The advantage this provides is that teams now become a virtual team. The teams will appear to operate "elbow to elbow" even though they seldom meet "face to face." The hypermedia aspect afforded by the Web and by Web Services makes this process doable today.

Emphasis on Learning: Given robust design algorithms, a hypermedia Functionbase described here will accelerate engineering such that more time can be spent finding ways to improve the design. Presently, many hours are spent rerunning codes, collecting data for presentations, and making reports, all of which could be automated. Time is redeemed for more erudite pursuits enabling the engineer to focus on improvements.

6.5 Summary of Knowledge Management

Computer and Information System technology has provided us with a new Knowledge Paradigm based upon the power to extract data and functions, such that the functional stranger can use it. It has also enabled new horizons for communicating new ideas and processes such that teams are becoming virtual. Historically, the customer and the various teams have been geographically remote, requiring the periodic design reviews that dictate a sequential design mode. Now, given this *Knowledge Management Paradigm*, the team can be extended and modified to include the customer, giving them insight into every aspect of design and implementation. Consensus, which was formerly developed post-priori using design reviews, can now be developed a priori using real-time communications of Functionbases.

Chapter 7
Agile, Robust Designs: Increasing Quality and Efficiency

Chain reaction–improve quality what happens? Your cost goes down. Half of the people here will understand that the other half will not… You can talk about quality, but if you don't know what to do about it, bring it about, quality is an empty work.
Mary Walton from one of W. Edwards Demming's Lectures.

7.1 Increasing Quality: Robust Designs

7.1.1 Defining Quality

Figure 5.6 implied quality which was a matter of balance. Another way to view this is to consider three principal elements in tension within engineering. They are People, Profit, and Process, illustrated in Fig. 7.1.

Conventional management is invariably focused on profit. This has been described by Walton [52] as a management disease in terms of: lack of consistency of purpose, emphasis on short-term profits, management by fear, and management mobility that creates prima donnas and dissolves commitment. All three of these concepts are antithetical to *Theory Z.*

Focusing on Process is the subject of the DoD Quality Master Plan, and is a strategy that is intended to drive continuous improvement at every stage of a program. Its objective is to combine management techniques, improvement techniques, and specialized tools, and a disciplined approach to process improvement. It states: "We have always managed the product to be in conformance to requirements, but we have not managed the processes that produce the products". Edward Deming [52] stated it this way: "do not manage the outcome; manage the activities that produce the outcome". This is consistent with a *Theory Z* organization in that team members must be disciplined to browse the Functionbases to identify improvements before the product is made: quality is not a priori. This is in contrast to the sign-off method where peer review is post-priori: the discipline of inspecting quality into the product seems easier, but it is always too little, too late.

Historically, the focus was on People through the use of formal organizations and mentorship. Young engineers are "mentored" by more experienced engineers, ensuring the continuance of product quality. The problem with mentoring is that

J. A. Crowder and S. Friess, *Systems Engineering Agile Design Methodologies*,
DOI: 10.1007/978-1-4614-6663-5_7, © Springer Science+Business Media New York 2013

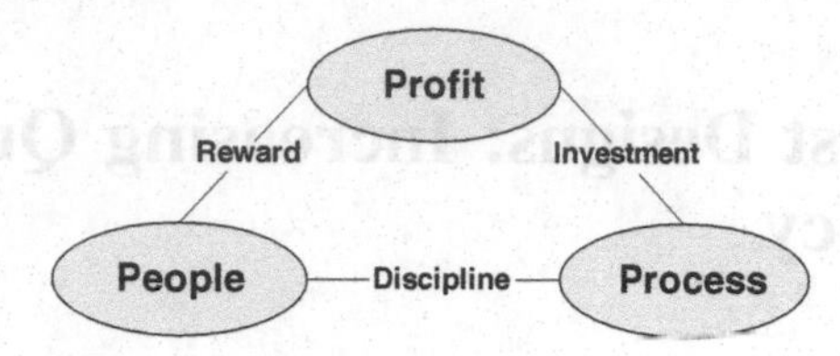

Fig. 7.1 Tensions that unravel quality engineering

the emphasis is on direct product knowledge and not on process knowledge. We do have processes for Systems and Software Engineering—but still these are not built into automated Functionbases that can be readily accessed within the contexts of product designs. They are stand-alone processes that we assume the engineers can fold into their everyday work. Automated tools that allow the processes to be easily folded into the design and implementation process from cradle to grave must accompany the processes.

The tension comes from wanting productivity: how many lines of code an hour can you produce? Not what kind of functionality can you provide with what quality, but how many lines of code did you write this month? The tension of quality versus speed is illustrated in Fig. 7.2. In comparison to mentorship, there is sometimes a push for a "quick and dirty" solution to a problem. And while this is necessary sometimes, most of the lessons learned through this are lost because the work is not captured adequately in an electronic Functionbase. And, because there is usually no reuse, it is wasteful and inherently lacks quality.

On the other hand, the Application-Specific Codes, which may be programmed into ASICS or FPGAs, are a good example of increased Quality and Speed, but

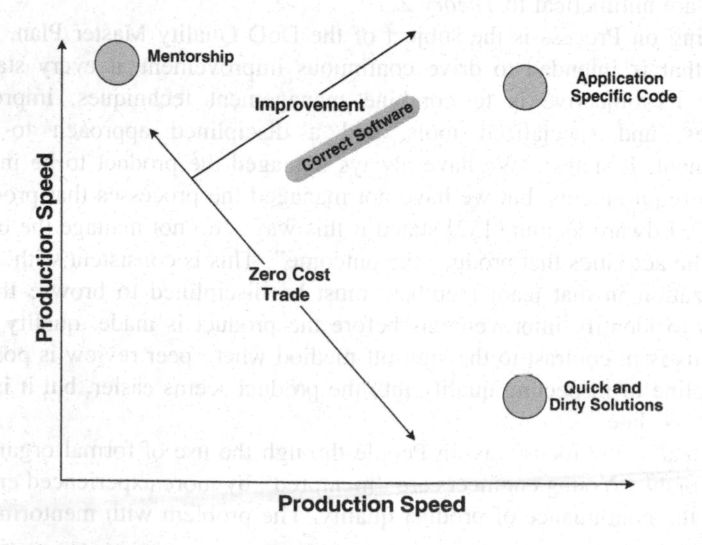

Fig. 7.2 Production quality versus production speed

may have very little reuse if their codes and designs are not captured in context with all of their information in one complete Functionbase.

Correct software is a challenge. Historically, programmers have depended upon fast compilers and debuggers. With faster computers it is tempting to depend extensively on the machine to debug code. However, "bug" is a euphemism for error and constitutes wasted effort. The objective must be to write provably correct code in order to both improve quality and reduce cost.

A statement often heard in design is, "if you want it really bad, you'll get it really bad", meaning that speed generally kills quality. Another similar statement is, "we don't have time to do it right, but we always have time to do it over". The conclusions would seem to be that you could have either quality or speed, but not both. This describes the tension between production speed and production quality. In either case, the common denominator is waste. Reduced waste can be correlated with both higher quality and higher speeds. To this end we have defined quality as a process–product dual.

Process Quality is achieved through the minimization of waste and "loss to society" through measurement and continual process improvement.

Product Quality is realized through the features and characteristics of a product, or service, which bear on its ability to meet and exceed user expectations (e.g., few discrepancies in our software).

7.1.2 Defining Performance Variance

As customers, we are disappointed when the products we have purchased do not perform the way we expected. On the other hand, we are delighted with it when it performs better than expected. The central theme of the QFD matrices is to define what is expected in terms of a set of **Quality Characteristics** that can be measured. Engineering analysis is then a matter of proving the extent to which the design exceeds expectations. The goal is to exceed, thereby reducing waste and delighting the customer [46].

This would appear to be a trivial exercise. However, it involves a change in culture: programs are driven by requirements, not expectations. The requirement is stated in terms of "goal posts", so that as long as the measured parameter falls within the posts, the requirement has been met. In contrast, the Quality Characteristic is something to be exceeded.

An example might be the quality of an interplanetary mission's trajectory. The customer demand forms the basic requirement, which is to send a payload to a neighboring planet. The engineering function might be on-board targeting to adjust the trajectory during the mission. The quality characteristic would be the minimization of normal energy (velocity perpendicular to that required). The customer also demands three bounds be placed on normal energy to ensure a minimum payload life once in orbit around the target planet. However, if quality means minimum waste, then the engineer is concerned with maximizing the payload's life expectancy. That means minimizing the variance of normal energy and exceeding the customer's expectations, not just conformance to a specified requirement. In short, "good is not good enough".

In a more general sense, there is requirements-pull and technology-push. The organization that responds only to requirements, works on the basis of requirements-pull; vision is nearsighted and management is by objectives only; conformance to specification is the predominant concept and quality is inspected into the product; technology is generated only to satisfy requirements and R&D groups must have funding approved by the program community.

The organization that responds to exceeding expectations is sensitive to technology-push; farsighted vision looks for new technology and concepts to integrate; products have quality because their processes have quality; R&D groups are looking to change paradigms from the outside of programs.

This was the farsighted goal of the Air Force's USAF R&D 2000 Variability Reduction Program (VRP) [35], illustrated in Fig. 7.3. Whereas most programs stop improving the design once conformance to the specifications is met, R&M 2000 calls for continual improvement, requiring the engineering process to exceed conformance [44]. In fact, the Statement of Work is to include a target for the "process capability index, *Cp*" defined as:

$$C_P = \frac{\text{Specification Range}}{\text{Process Range } (6\sigma)}$$

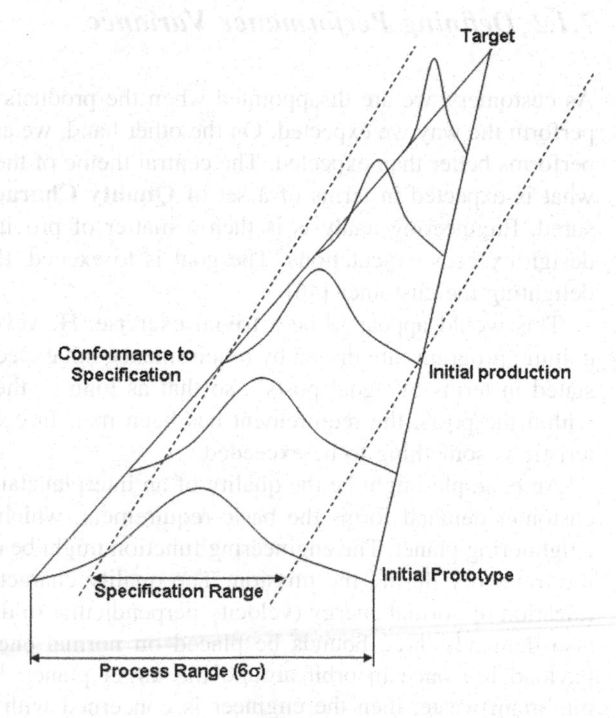

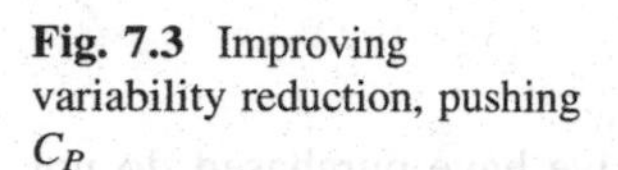
Fig. 7.3 Improving variability reduction, pushing C_P

where the specification range is a minimum performance requirement and the Process Range is the performance measured by test or, more likely, by Monte Carlo simulation.

The objective would be to drive *Cp* to be greater than unity, such that the system behavior is easily predicted and every item is identical. Dr. Demming [52] describes this in terms of comparing his hometown orchestra to the London Symphony Orchestra, "Same music; same specifications. There are no mistakes; both play the right notes, but listen to the difference, just listen to the difference".

7.1.3 Statistical Process Control

With the variance computed and the sensitive parameters identified, SPC can be applied for program management to remotely monitor the Quality Characteristics identified during the QFD process development. Not all parameters will be identified for reduction; only some may simply be present [6]. It is important to note that because the parameter under statistical control is a moving target (should be changing), it does not mean the process is out of control. As has been explained, engineering design means taking an idea, implementing it, then turning it into reality. There is always a lot to be learned along the way, where experimental data are accrued and lessons learned. We are never certain about the outcome of our assumptions at the beginning of the design. "How do I know what I think until I see what I say", is one way of describing this, a line attributed to the Harvard psychologist Jerome Bruner. The purpose is to get the design moving, focusing on the most important objectives determined during the QFD process.

7.2 The Software Cleanroom

7.2.1 The Cleanroom Concept

The *Software Cleanroom* is a concept that integrates the product teams, information technology, and the demand for increasing quality. It is designed to fulfill the long-term vision: to develop and integrate new technologies into a software development program. The quality goal is to increase the engineer's productivity tenfold over time and make continual improvement a reality.

Cleanroom software engineering is a concept that integrates statistical quality control into engineering. There are three attributes defined for the cleanroom discipline:

1. A design methodology that prevents defects in preference to correcting them;
2. A design that is incremental in preference to sequential; and

3. Software quality that is measured through representative use in preference to counting discrepancies.

The first attribute is provided through the engineering tool-case tool symbiosis. The second is achieved through the integration of the QFD matrices, the sandbox, and the target system. Monte Carlo simulations and a subset, the Use Cases, run on the target system, provide the third attribute.

7.2.2 VRP Requires the Discipline of a Cleanroom

The cradle-to-grave view is illustrated in Fig. 7.4. In addition to the software team, there is an integration team and the test (e.g., V&V) team. The objective is for the software product teams to design, build, and test their software remotely from the target system. The Test team validates the implementation using the identical Quality Characteristics drawn from the QFD database and the Use Cases. The quality of the code is measured against the proportion of the Use Cases, or User Scenarios, the implementation passes. This form of quality measurement has little relation to system reliability and dependability of the target system code. Based on the User Test results, process improvement requirements can be identified.

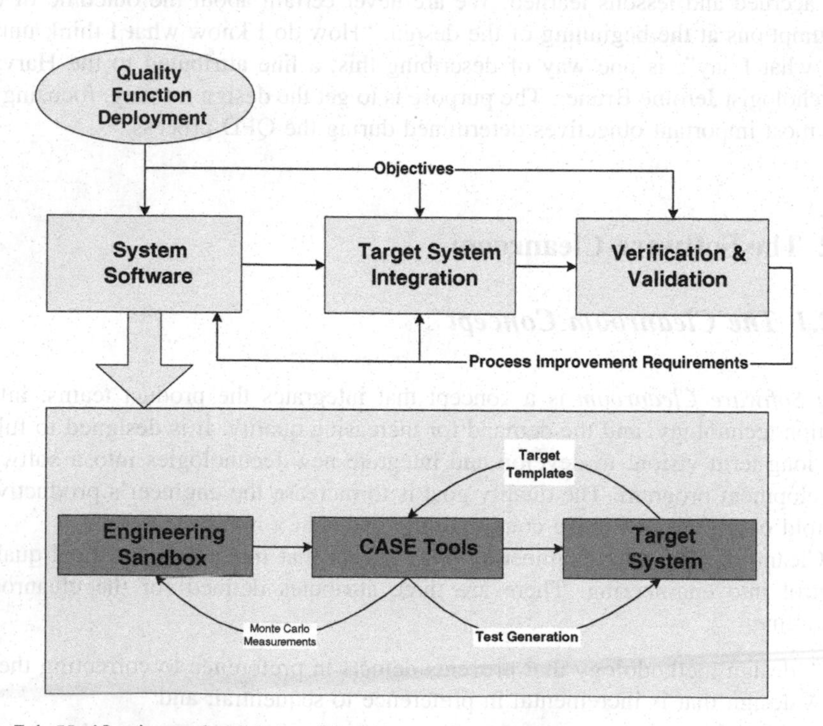

Fig. 7.4 Verification and validation drives process improvement

For example, if the User Tests reveal timing frames are being "blown", the required improvement will be to the Resource Architecture Library, referenced in Fig. 6.7.

Discipline in the cleanroom will result in corrections being made to the engineering process, not the product. This ensures the lesson is truly learned and that the error is not repeated.

7.2.3 Stable Specifications and Certification

The QFD A1 matrix is the cornerstone of the cleanroom and constitutes the first increment in the design process. The A1 and A2 combined provide the bases for measuring design progress through the Quality Characteristics. Using the reusable libraries in the Functionbases, the software components in the A4 matrix can change rapidly. Progress is not measured relative to the number of A4 components completed nor to the lines of code produced. Process is monitored using *Cp*. Deriving stable A1–A2 matrices becomes important as it enables clear observation about the quality of the software process. Without it, low process quality will be lost in the noise caused by ever-changing specifications.

The objective of the software cleanroom is to be able to certify the system software. This is achieved by using SPC to monitor the Quality Characteristics and Use Case statistics. Certification is achieved once conformance has been proved (i.e., $Cp = 1$). Once the software has been qualified, it can be placed in the Functionbase reuse library, along with all the relative materials (e.g., designs, requirements, etc.) that compose a Functionbase.

7.3 Modern Systems/Software Development Tools

Much of the processes discussed here are facilitated through Commercial Off-the-Shelf (COTS) software products that are currently available. There are several versions of the Electronic Engineering Notebook available. One is the E-WorkBook Suite © by IDBS, The Electronic Lab Notebook © by LabArchives, and the Oak Ridge National Laboratories (ORNL) Electronic Engineering Notebook Project, sponsored by the DOE 2000 Electronic Notebook Project.[1] The purpose of these products is to provide an electronic equivalent to the paper research or lab notebooks engineers have used for many decades. It will record sketches, equations, plots, graphs, images, signatures; everything during the process from R&D, to Systems Design, to Software Design and Development, through Integration and Test, through maintenance, i.e., cradle-to-grave.

[1] http://www.csm.ornl.gov/~geist/java/applets/enote/#demo

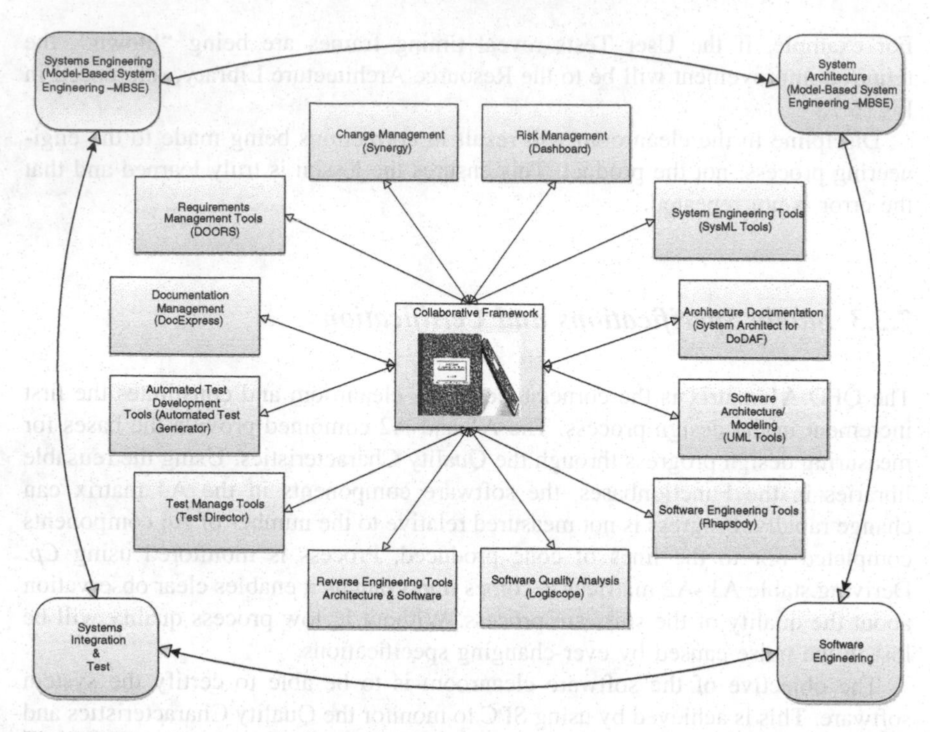

Fig. 7.5 The agile systems/software development process with reverse engineering comparisons

The use of modern software development tools allows the Systems and Software Architectures to be linked to requirements. This then allows the architecture to be linked to class diagrams and then ultimately to the code to provide the mechanisms to create the software "cleanroom" described in this book. By reverse engineering the code in a "cleanroom" and then comparing the resultant software design (based solely on the code), a comparison can be made to determine if the architecture of the code written has resemblance to the original software architecture, i.e., "does the code do what it was architected to do"? Figure 7.5 illustrates this process.

Architecture development tools, such as Rhapsody®, Control Center®, and many others, in conjunction with the Electronic Engineering Notebook, Functionbases and cleanroom concepts, provide the capability to forward and reverse engineer the code for comparisons. This provides the necessary validation that the code meets requirements by verifying that the code written matches the architecture and ties to the requirements that drove the architecture.

All of this work would be captured in the Electronic Engineer's Notebook so that the entire process and Functionbases can be delivered and archived for future software and architecture reuse.

Chapter 8
Conclusion: Modern Design Methodologies—Information and Knowledge Management

An expert is someone who knows some of the worst mistakes that can be made in his subject and how to avoid them.
Werner Heisenberg, Der Treil und das Ganze, Augarde [1].

8.1 Embracing Change: Everyone Should be an Expert in Change

People dislike change. It takes them out of their comfort zone. Human nature demands continuity and consistency. These factors must be taken into consideration in our engineering if continual improvement is to be realized. There are telltale signs that indicate when change is a natural part of the process:

Quality is measured to prove the customer's issues are being met and to provide the basis for forcing change. Changes required for the Continual Process–Product Improvement (CPPI) become an everyday affair.

Quality is implicit and is as much a part of the engineering as correct math; if the team cannot generate correct math, then the organization needs a math department. The same is true of quality. So if the organization has a quality department, *it indicates engineering processes that are not robust cannot integrate change*.

Quality is related to training and education because change is perpetual. Quality is a race to be run, not just an objective to complete. Therefore, it is likely that at least one-fifth of an engineer's time will be spent in training and education. Without it, the Interactive Experiments required to find improvement opportunities will have low yields.

A Theory Z organization will ensure the team owns the product and the resources. In some respects, this is a return to Taylor's Scientific Management where the team is the industrial engineer, determining how the resources can best be utilized for process–product improvement [48]. These resources may be allocated to education, faster hardware and software, additional team members, etc. This may require an organizational focus on long-term profit before commitment can be made, as the cultural change involves the engineer as well as the manager. Some short-term losses will no doubt be incurred as the team learns to take advantage of the new paradigm.

J. A. Crowder and S. Friess, *Systems Engineering Agile Design Methodologies*,
DOI: 10.1007/978-1-4614-6663-5_8, © Springer Science+Business Media New York 2013

8.2 Robust Designs the First Time

The Expert Systems Designer is our concept generated by all the ideas discussed in this chapter and is graphically represented in Fig. 8.1.

8.2.1 The Expert Systems Designer

The attributes of the Expert System Designer are described as follows:

Quality Function Deployment (QFD) provides a means of capturing an organization's existing knowledge for reuse by less experienced engineers. In this manner, less time is wasted generating the product that best meets the Customer Demands as old mistakes are not repeated and the wheel is not redesigned. This could even be taken to the point of using Functionbase interrogation to find the best QFD to fit a new set of Customer Demands, or even amalgamating components of several QFD Functionbases to forge an even better fit. All that would then remain is to check that the design is *Right-First-Time* through the performance estimation.

Provable Synthesis Methods and Tools provide the foundation for design without error. The fact that most errors are repeated is not new; this is why code

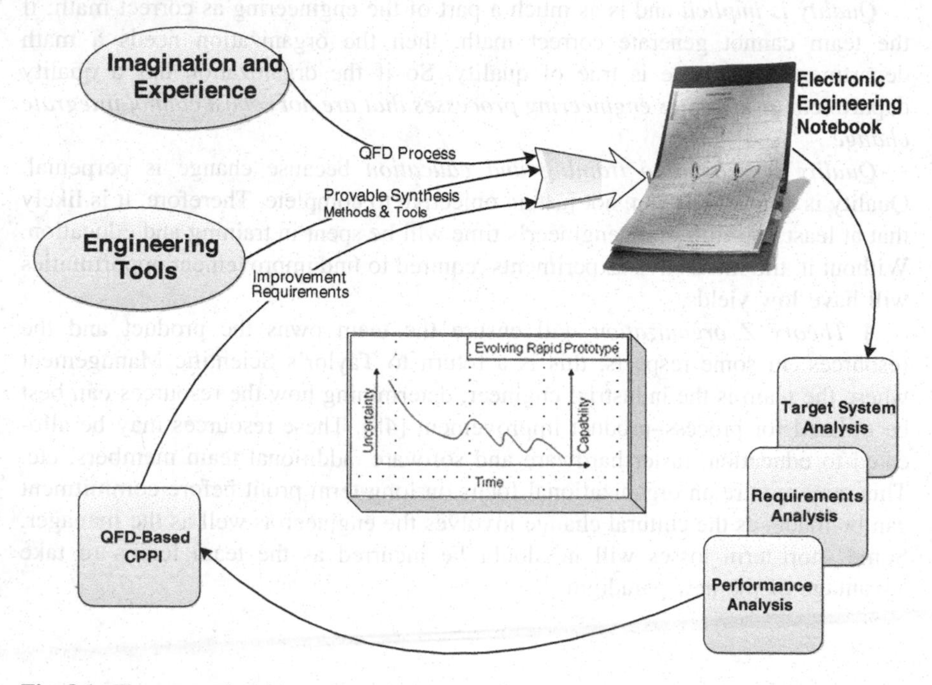

Fig. 8.1 The expert designer: agile systems engineering

libraries have been made for reuse. What is required is to integrate the knowledge about repeated errors to affect robust error checking, hence the Case Tools. Also, there is the question of the quality of the requirements. This requires the analyst to question his own thinking through experimentation. The Engineering Tools provides the Engineering Sandbox with rich function libraries. The two tools together provide a synergism where new ideas can be developed and implemented *Right-First-Time*. The method for eliminating errors is aggressively preventative in nature, not corrective as post-priori bug elimination is wasteful and uncertain.

The Electronic Engineer's Notebook integrates the QFD, methods-tools, and improvement requirements. It enables the functional-stranger to rapidly repeat an analysis or regenerate a design. It is the Notebook that will enable the engineer to spend less time on clerical work and, through automation, more time on invention and design.

Analysis of Performance estimates, form a part of the *Electronic Engineering Notebook* (EEN). Development of models provides the means to test and tighten performance estimates and validate what we think we know. This provides the basis for change; changing what we don't know is always dicey. With proven knowledge, change can be made in a managed fashion.

QFD-Based Continual Process–Product Improvement (*CPPI*) is the integration of new technology, concepts, and knowledge. Decision-making is based on *Cp* measurements and how they will be affected. Such a rational basis is a prerequisite to paradigm busting. QFD is a discipline that ensures the product does not suffer paradigm paralysis. If it dies, the team will experience King Harold's folly.

The Evolutionary Rapid Prototype describes the capability realized by the Expert Designer concept. Evolution is a gradual process in which something changes into a different, and usually more complex, form. Rapid means moving and moving swiftly. Prototype is an original type, form, or instance that serves as a model on which later stages are based or judged. Therefore, the Evolutionary Rapid Prototype means each process–product is a basis for the next. The process–product is ever changing and always improving.

8.2.2 Continuous Improvement and Technology Revolution

Three of the Required Engineering Domains described in Fig. 6.8 are shown in Fig. 8.2 to illustrate their relationship to each other. The engineer's time might well be split evenly between each domain to generate the necessary rate of improvement. Treating any one of the domains in isolation will result in ineffectual change. Also, loss of balance between the three will result in loss of competitiveness. Lack of commitment to the three by the organization is a lack of commitment to Quality, People, and Process.

Recurring Engineering in enabled through the EEN, and allows software loads to be generated and tested automatically [40]. With all the functions predetermined

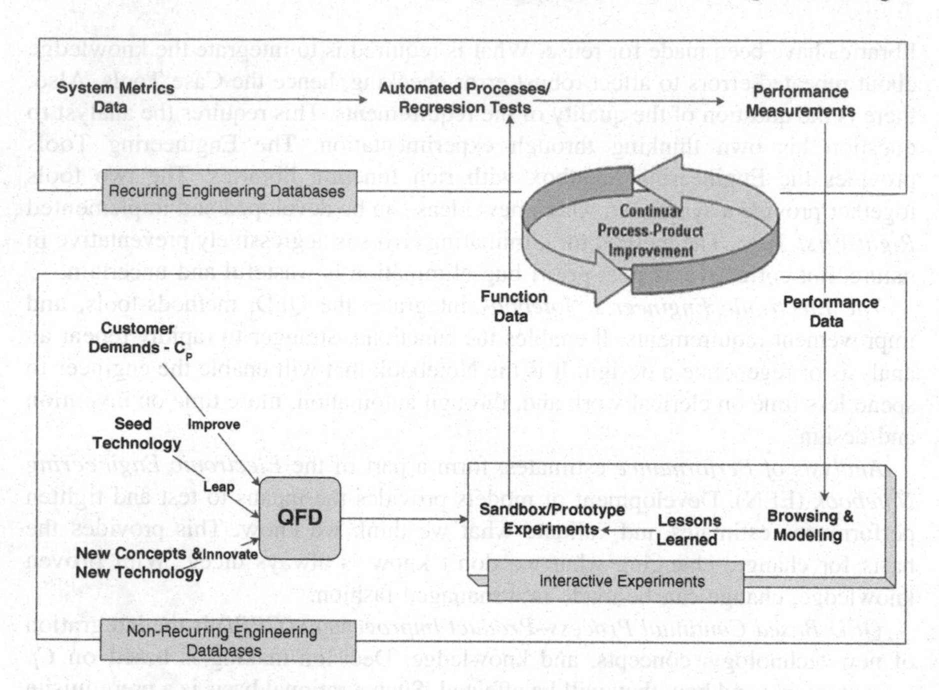

Fig. 8.2 Continual improvement requires an integrated approach to engineering

and all the design tools interfaced through Knowledge Management, the generation of the software loads is simply a matter of CPU time. The process is as follows:

- The analyst completes the data definitions using the data management tools within the EEN.
- Once complete, the engineer can invoke the design macros built into the EEN through the Knowledge Management Process. These macros know how to read the data and spawn the design processes.
- These processes include the Monte Carlo simulations and/or regression tests (if required) to measure performance; therefore performance analysis is built into the process that enables requirements verification.

Interactive Experiments are essential to CPPI as it indicates where the paradigm is weak or broken. Given robust error checking methods and thorough knowledge captured through QFD, there still remains sets of errors that can be traced to inadequate requirements analysis. When addressing this domain, there does not appear to be any substitute for human intuition and insight. At present, the organization depends on peer reviews using viewgraphs and questions. The Expert Designer makes the process–product Functionbases available to anyone over the internal organization web through a variety of collaborative environments, thereby providing familiarity normally reserved for the designer, to the whole community. The least member of the technical community will use the Recurring Engineering functionality and the erudite will use the Interactive Experiments components.

Non-Recurring Engineering has three drivers: improvements in *Cp*, leaping to new products (e.g., the Evolving Rapid Prototype), and innovation through new technology and concepts [19]. Making the Recurring Engineering, and Interactive Experiments Functionbases, available to the community will generate an abundance of ideas. The QFD process will manage all these. Each idea to be evaluated against Customer Demands and the impact on Quality Characteristics, then prioritized for new engineering. Discrete engineering reviews will become outmode as the SPC provides progress data real-time. Even the Test Readiness Reviews are overcome by the Software Cleanroom concept for software quality as measurement criteria are established using QFD on day one: if the Recurring Engineering measurements fail minimum requirements, the design in not ready to fly [43].

QFD is employed throughout the life cycle to manage the improvements. Because paradigm paralysis is such a de-motivator to change, an objective basis must be forged to break the deadlock. The QFD process integrates all of the necessary considerations; cost, risk, quality, etc. The list of improvements can be ordered such that the team's resources are always focused on the largest payoff. Diminishing payoff identifies the need for a new paradigm (i.e., new technology and R&D).

CPPI is a consequence of implementing this paradigm. Quality is not achieved through reacting to failures, it requires both the time and desire to improve the product: automated processes are a prerequisite if time is to be redeemed for new designs; performance measurement is mandatory if quality is to be more than a figment of the manager's imagination; browsing and modeling are essential for generation of new ideas and validating existing knowledge; sandbox experiments are the basis for new and novel designs. The whole process is one of "to do, then discover". Improvement and technology development are perpetual.

8.2.3 Agile Systems Engineering: A Holistic View

Great Spirits have always encountered violent opposition from mediocre minds.

Albert Einstein

Making change an integral component of a design methodology affects the whole product and process life cycle. Because the engineer and his tools are part of this, and because computer technology has had such a widespread impact, a holistic view must be taken. This chapter is not exhaustive in its coverage of the material, but has sought to address the major aspects.

The New Organization is a focus on how change in engineering requires change in work organization. Engineering organizations were designed when engineering methods were manual and unsophisticated. This new paradigm is possible through the use of modern Information Systems technology and Web Services to manage the mass of information. The real question to be answered is:

will the manager cede control to the team? The solution will most likely come through economic necessity, and the sheer inevitability of the Information Age. Whatever happens, a valuable tool in the arsenal is QFD. Engineers can use it to build technology roadmaps to new products. Managers can use it to build the New Organization to improve Customer Satisfaction.

Increasing Quality should be closely linked to Imagineering. Imagine a process with all waste eliminated. Now design it. In contrast, engineering that is driven by conformance-to-spec is suffering paradigm paralysis. To eliminate waste there must be quality in the process, as this will affect quality of the product. For the design engineer, quality and robustness are often linked, as the measure of performance can be the same for each. Therefore, advanced design algorithms have become even more valued as a means of reducing variance. The toughest challenge for the design engineer may be to make statistics an intuitive skill; solutions to improving quality will then follow more comfortably [23].

Finally, there are two major points that this chapter is trying to put across:

- Quality and robustness are defined through measurements, and maintained through automated measurements (e.g., regression tests), not impressions; and
- Higher quality demands change, which will happen with or without us.

References

1. Augarde, T. (1991). *The Oxford Dictionary of Modern Quotations*. Oxford: Oxford University Press.
2. Azim, S., Gale, A., Lawlor-Wright, T., Kirkham, R., Khan, A., & Alam, M. (2010). The importance of soft skills in complex projects. *International Journal of Managing Projects in Business, 3*(3), 387–401. doi:10.1108/17538371011056048.
3. Barker, J. (2000). *Discovering the future*. New York: Harper Collins.
4. Beck, K., et al. (2001). Manifesto for agile software development. Agile alliance.
5. Beck, K. (2002). *Test driven development*. Boston, MA: Addison-Wesley Professional.
6. Bendat, J. (2000). *Non-linear system analysis and identification*. New York: Wiley Interscience.
7. Black, N. (2011). Technology changing at lightning fast speeds. *The Daily Record* (Sept. 21).
8. Boeree, C. (2006). *Alfred Adler: Personality theories*. Shippensburg, Pennsylvania: Shippensburg University.
9. Books, F. (1975). *The mythical man month*. Boston, MA: Addison Wesley.
10. Burns, R. (1996). Paper comes to the West, 800–1400. In *Lindgren, Uta, Europäische Technik im Mittelalter*. 800 bis 1400. Tradition und Innovation (4th ed., pp. 413–422). Berlin: Gebr. Mann Verlag. ISBN 3-7861-1748-9.
11. Checkland, P. (2001). *Systems thinking, systems practice*. Hoboken, NJ: Wiley.
12. Clausing, D. (2001). Concurrent engineering. *Design and Productivity International Conference*.
13. Conway, W. (1995). *The quality secret: The right way to manage*. Nashua, NH: Conway Quality.
14. Crawford, B., Leon de la Barra, C., Soto, R., & Monfroy, E. 2012. Agile software engineering as creative work. In *Proceedings of the 5thInternational Workshop on Co-operative and Human Aspects of Software Engineering*, ICSE, Zurick, Switzerland.
15. Crowder, J. A. (1996a). RLV mission system architecture. *NASA X-33 Report, 96-RLV-1.4.5.5-004*, Littleton, CO: Lockheed Martin.
16. Crowder, J. (1996b). Mission planning and system operations life-cycle concepts. *NASA X-33 Report, 96-RLV-1.4.5.5-006*, Littleton, CO: Lockheed Martin.
17. Crowder, J. (1993). *Making change an integral component of advanced design methodologies*. San Diego, CA: Academic.
18. Crowder, J. (1997). Stochastic models for rapid assessment of complex, deterministic engineering systems. *Dissertation, La Salle University*.
19. Crowder, J. (2001). Integrating metrics with qualitative temporal reasoning for constraint-based expert systems. *NSA Technical Paper—CON-SP-0014-2002-08*.
20. Crowder, J. (2002). Flexible object architecture for the evolving, life-like yielding, symbiotic environment (ELYSE). *NSA Technical Paper—CON-SP-0014-2002-09*.
21. Crowder, J. (2003a). Using a large linguistic ontology for network-based retrieval of object-oriented components. *NSA Technical Paper—CON-SP-0014-2003-03*.

J. A. Crowder and S. Friess, *Systems Engineering Agile Design Methodologies*,
DOI: 10.1007/978-1-4614-6663-5, © Springer Science+Business Media New York 2013

22. Crowder, J. (2003b). Ontology-based knowledge management. *NSA Technical Paper—CON-SP-0014-2003-05*.
23. Crowder, J. (2003c). Agile business rule processing. *NSA Technical Paper—CON-SP-0014-2003-06*.
24. Davidson, J. (1991). *The great reckoning*. Orangeville, Ontario: Summit Books.
25. Dierolf, D., & Richter, K. (2000). Concurrent engineering teams. *Technical Report*, Institute for Defense Analysis.
26. Drucker, P. (1998). The coming of the new organization. Boston, MA: Harvard Business Review.
27. Gillespie, R. (1991). *Manufacturing knowledge: A history of the Hawthorne experiments*. New York: Press Syndicate of the University of Cambridge.
28. Hamilton, M., & Hackler, W. (2000). A rapid development approach for rapid prototyping based on a system that supports its own life cycle. In *Proceedings, 10th International Workshop on Rapid System Prototyping*.
29. Heinz, J. (2000). What went wrong. *Aerospace and Defense Science*.
30. Hershey, P., & Blanchard, K. (1989). *Management of organizational behavior: Utilizing human resources*. Saddle River, NJ: Prentice-Hall.
31. Honald, L. (1997). A review of the literature on employee empowerment. *Empowerment in Organizations, 5*(4), 202–212.
32. Hostyn, J. (2012). Designing change: Sensemaking in a changing world. www.joycehostyn.com/blog. 16 Sept 2012.
33. Humphrey, W., & Sweet, W. (2007). *A method for assessing the software engineering capability of contractors*. Carnegie Mellon: Carnegie Mellon.
34. Jacobs, E., Mason, R., & Harvill, R. (2002). Group counseling strategies and skills (4th ed.). Pacific Grove, CA: Brooks/Cole.
35. Johnson, B. (2001). R&M 2000 variability reduction process. *Technical Report*.
36. Johnson, C. "Kelley." (1985). Kelly, more than my share of it all. Washington, DC: Smithsonian Institute.
37. King, B. (1997). Better designs in half the time. Methuen, MA: GOAL/QPC.
38. King, B. (1992). *Better Designs in Half the Time: Implementing QFD Quality Function Deployment in America*. Methuen, MA: GOAL/QPC.
39. Lawson, M. (2008). The battle of hastings, 1066. Indiana: Indiana University.
40. Lucena, J. (2006). Special issue globalization and its impact on engineering education and research. *European Journal of Engineering Education, 31*(3), 321–338.
41. McHugh, O., Conboy, K., & Lang, M. (2012). Agile practices: The impact on trust in software project teams. *IEEE Software May/June* 2012.
42. Menker, L. (2000). Results of the aeronautical systems division critical process team on integrated product development. *Technical Report*, Wright Patterson AFB.
43. Mills, D. (2002). Cleanroom software engineering. *IEEE Software*.
44. Papoulis, A. (1965). *Probability, random variables, and stochastic processes*. New York: McGraw Hill.
45. Rowley, D. (2007). *Forming to performing: The evolution of an agile team*. Washington, D.C.: IEEE Computer Society.
46. Safonov, M. & Laub, A. (2001). The role and use of the QFD matrix. *IEEE Transactions in Automatic Control*.
47. Tansik, D. (2000). Technological gatekeeping. *Technical Report*. Arizona: University of Arizona.
48. Taylor, F. (1911). *Shop management, the principles of scientific management*. New York: Harper and Row.
49. The Bible, Joseph and Pharaoh; agricultural control. Genesis 47:23.
50. The Bible, Old testament bureaucracy. Exodus 17:18.
51. Tuckman, B. (1965). Developmental sequence in small groups. *Psychological Bulletin 63*(6), 384. PsycArticles-1 June 1965.

52. Walton, M. (1986). *The Deming management method*. Mesa, AZ: Perigree Books.
53. Wan, T., Zhou, Y. (2011). Effectiveness of psychological empowerment model in the science and technology innovation team and mechanism research. In *Proceedings of the 2011 International Conference on Information Management, Innovation Management, and Industrial Engineering* (ICIII).
54. Weber, M. (1947). *The theory of social and economic organization*. New York: Simon & Schuster Inc.
55. Young, M. (2009). A meta model of change. *Journal of Organizational Change Management, 22*(5), 524–548.

Index

J. A. Crowder and S. Friess, *Systems Engineering Agile Design Methodologies*,
DOI: 10.1007/978-1-4614-6663-5, © Springer Science+Business Media New York 2013

Printed in the United States
By Bookmasters

Printed in the United States
By Bookmasters